Making
SIMPLE AUTOMATA

Making
SIMPLE AUTOMATA

Robert Race

THE CROWOOD PRESS

First published in 2014 by
The Crowood Press Ltd
Ramsbury, Marlborough
Wiltshire SN8 2HR

enquiries@crowood.com

www.crowood.com

This impression 2024

British Library Cataloguing-in-Publication Data
A catalogue record for this book is available from the British Library.
ISBN 978 1 84797 744 1

Frontispiece: *The Motley Crew.*

Picture credits
Page 8 ©Rijksmuseum van Oudheden, Leiden, NL.
Page 9 (*bottom*) Library of Congress, prints and photographs collection. LC-USZ62-110278.
Page 10 Los Angeles County Museum of Arts.
Page 20 D. J. Shin, CC BY-SA3.0

ACKNOWLEDGEMENTS

I am grateful to all the makers, named in the text, who have generously allowed me to use pictures of their work. Many thanks to them, and thanks also to the anonymous makers of the simple, but ingenious, mechanical toys that so inspire me, and special thanks, for all her support, to my wife, and favourite Muse, Thalia.

Typeset by Kelly-Anne Levey
Printed and bound in India by Parksons Graphics Pvt. Ltd.

CONTENTS

INTRODUCTION

WHAT ARE AUTOMATA?

Automata is the plural of the Greek word *automaton*, meaning a thing that moves of itself. The plural can also be automatons, but it is less common. Rather more common, but not strictly correct, is the use of automata for the singular. Modern dictionaries give a broad spectrum of definitions and usage. Sometimes the word has a narrow specialized application, such as in the mathematical concept of a cellular automaton. It can be applied to a person, or to a living creature in general, when it suggests a possibly efficient, but merely mechanical action, without thought or feeling. One of the common usages of the word focuses on the notion of a mechanical device that moves, is usually intended as a toy or amusement, and often imitates the action of a living creature. There may be reference to a concealed mechanism and motive power. Although automata can be quite complicated machines, mimicking the movements of human beings or animals, even performing complex actions such as drawing a picture or writing, the Oxford English Dictionary gives a clockwork mouse as an example.

This book deals with the design and construction of small scale, simple mechanical devices made for fun. I shall call them automata, although in many, the mechanism, rather than being *concealed,* is in full view and intended to be part of the overall effect. The source of motive power and its transmission are also often clearly visible. Power may be provided directly by turning a drive shaft with a crank handle, or less directly, for instance by raising a weight or winding a spring.

THE HISTORY OF AUTOMATA

That the OED gives a clockwork mouse as an example of an automaton indicates that it is pretty impossible to disentangle the history of automata from the history of moving or mechanical toys, not to mention that of puppets and dolls, of kinetic sculpture, of theatrical devices, of conjuring or of robotics.

If you look at such histories you will find the same passages from classical authors, and the same museum objects, claimed as early references to, and early examples of, automata, or of moving toys, or of dolls, or of puppets and so on.

OPPOSITE PAGE: *Canoe with Birds.* **A simple automaton, made by the author in driftwood.**

RIGHT: **The Oxford English Dictionary suggests a clockwork mouse as an example of an automaton.**

Figure Kneading dough. An ancient moving toy, operated by pulling a string. Egypt, around 2000BCE.

It is well worth following these histories. Since prehistoric times the urge to represent living things by animating them has been a significant factor in the development of technology.

Among the toys, or toy-like objects, found in ancient Egyptian tombs are a variety of jointed figures with movable limbs, and animals with moving jaws, operated by pulling a string. One notable example that survives in the Egyptian Museum in Cairo is an ivory sculpture of three dancing dwarves mounted on a base: strings can be pulled to rotate the figures (a fourth figure from the group is in the Metropolitan Museum, New York).

A number of moving toys, including wheeled toys, dolls with movable limbs and figures operated by pulling a string, survive from ancient Greece, and also from earlier civilizations, such as those in Mesopotamia and the Indus valley.

References to automata, moving toys and puppets in classical texts are sparse, and often difficult to interpret with any certainty, but they do suggest that such things were familiar objects in ancient Greece. For example, in Chapter 7 of *The Republic,* in the well-known allegory of the cave, Plato pictured puppets, with their operators hidden behind a wall below, casting shadows on the wall. Aristotle, in *De motu animalium,* compared the movement of animal limbs to automata (by which he probably meant some sort of puppet, or mechanical theatrical device) and also to a strange-sounding toy cart, which runs in a circle because the wheel, or wheels, on one side are smaller than on the other.

Solid descriptions of more complex automata first appear in the second and first centuries BCE in the works of the Alexandrian school, and particularly of Ctesibius, Philo and Hero. A wide variety of hydraulic, pneumatic and mechanical devices are described in the texts that survived. Some of these, such as clepsydrae (water-clocks), are machines with a really practical purpose, but they often incorporate moving figures, singing birds and, literally, bells and whistles. In addition, many are really just automata – intended to mystify theatre-goers and temple-goers, or simply to amuse.

These texts survived because they were transcribed by Byzantine and Arab scholars. They subsequently had a considerable impact on Renaissance Europe when Latin translations appeared during the sixteenth and seventeenth centuries.

Woman with Rolling Pin, by a modern maker, Edessia Aghajanian. In the British Museum there is a terracotta string-pull toy from Rhodes, dating from 450BCE, showing the same timeless domestic task.

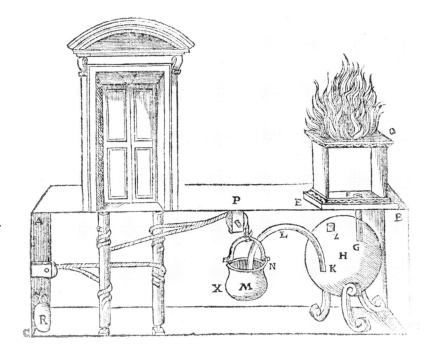

Hero's device for automatically opening temple doors when a fire is lit on the altar. From Giambattista Aleotti's work of 1647.

Under Harun al-Rashid, the fifth caliph of the Abbasid dynasty, who succeeded in 786CE, and under his successors, Baghdad became an important centre of learning, not least in mathematics and science. Scholars at the House of Wisdom actively sought Greek texts, such as Euclid's *Elements*, to translate into Arabic. Among them were the three Banū Mūsā brothers who produced the *Book of Ingenious Devices* including an array of hydraulic and pneumatic automata including trick vessels, automatic fountains and music machines. They drew heavily on the work of Philo and Hero, but introduced many original features of their own.

Another important figure in the history of automata is the brilliant engineer Al-Jazari, who, in the late twelfth century was in the service of three successive Artuqid rulers in the city of Āmid, now Diyarbakir in Eastern Turkey. His amazing *Book of Knowledge of Ingenious Mechan-*

Al-Jazari's design for a candle clock, with a system of weights and pulleys, releasing twelve balls to mark the hours.

ical Devices appeared in 1206CE. It also draws heavily on the Alexandrians, and on the Banū Mūsā brothers' developments. It has numerous coloured drawings of automata and other hydraulic, pneumatic and mechanical constructions, many involving delicate control mechanisms, and highly sophisticated water clocks and a boat with four mechanical musicians operated through a camshaft. The drawings are detailed, if sometimes difficult to interpret. Al-Jazari's designs use mechanical devices, such as crankshafts and camshafts, and are technologically advanced in the use of segmental gears, and of conical valves with the seats and plugs ground down to give a watertight fit.

Grottoes, fountains and mechanical theatres

Translations of Hero and Philo started to appear in Europe at the beginning of the sixteenth century. The hydraulic and pneumatic devices that they described were used enthusiastically in princely pleasure gardens with ever more elaborate fountains and grottoes with mechanical music, moving figures and automated scenes. The motive power was provided by flowing water, but the mechanical elements became more and more complex, using pulley systems, camshafts and crankshafts to animate individual figures and even whole theatrical scenes.

These garden embellishments were the preserve of the seriously rich and powerful – they were expensive to build and difficult to maintain. Many of the more intricate mechanical elements did not last. However, a famous example of these waterworks was installed early in the seventeenth century at Schloss Heilbrunn near Salzberg and many of the effects survive, although the hydraulic mechanisms are not original. In the eighteenth century the Nuremberg craftsman Lorenz Rosenegge installed an extraordinary model theatre, which is also still in working order.

Lorenz Rosenegge's water-driven mechanical theatre added to the attractions in the gardens at Schloss Hellbrunn, Salzberg, in the mid-eighteenth century.

Clockwork

In 807 one of the gifts that Harun al-Rashid presented to Charlemagne was a clepsydra (water clock) in which twelve horsemen appear in turn and twelve balls fall onto cymbals to strike the hours. There were plenty of other opportunities for the advancing technologies used in clepsydrae (water clocks) in the Arab world to pass into Europe, principally through Muslim Spain. Complicated water clocks became widely used in medieval Europe, and usually they would include various automata, such as figures to strike a bell to mark the hours. From the end of the thirteenth century weight driven *clocks*, with an escapement mechanism to control the fall of the weight, started to replace them. The word clock comes from the French *cloche*, Latin *clocca*, meaning *bell*, and the earliest had no hands, indicating the passage of time by the ringing of

a bell. The principal purpose, for these mechanical clocks, as for the clepsydrae that preceded them, was to measure the fixed canonical hours of prayer. They were substantial structures, and typically mounted on a tower. Just as the makers of clepsydrae often incorporated animated figures, opening doors and singing birds, the mechanical clock makers incorporated animated strikers of the bell to be operated by the weight-driven clockwork mechanism. These were known as jacks of the clock, or jaquemarts.

By the late fourteenth century dials were being added to the clocks. At first the dial rotated and a fixed hand indicated the time, but soon a rotating hand sweeping around a fixed dial became the norm. Mechanical clocks developed rapidly. There would be more than one jack to strike the bell, and different sets of figures might appear through doors at the hour and at the quarters, circling around at the front. During the fifteenth

Wells Cathedral. Jack Blandiver

A jack of the clock in Wells Cathedral. The figure strikes the bell in front with a hammer, and two bells with his heels.

and sixteenth centuries many churches also had automata associated with the organ, and mechanical figures of Christ on the cross, or of devils and angels, were not uncommon.

Gradually the mechanisms of large public clocks were refined and miniaturized, as were the automated figures associated with them. During the sixteenth century, notably in southern Germany, increasingly sophisticated animated human figures, *androids,* and devices on a domestic scale, such as the table decorations based on the *nef* were constructed. The nef was a model ship, originally for storing utensils and spices, but it developed into an elaborate mechanical ornament incorporating a clock, with many little automated figures.

By the eighteenth century there was a growing appetite throughout Europe for ever more life-like and active mechanical figures, powered by more complex clockwork, with springs, weights, gear trains, wires, chains, pulleys, bellows, cranks, camshafts and levers. In 1738 Jacques de Vaucanson, a brilliant mechanician and astute showman, exhibited *Le Flûteur* (The Flute Player) in Paris. This was an extraordinary life-size figure, mounted on a plinth that concealed the complicated mechanism which controlled the flow of air, changed the shape of the mouth, moved the steel tongue, and operated the fingers. It could play twelve melodies. The following year he added two more automata: the first was a figure that played the traditional Provençal combo of *galoubet et tambourine* (pipe and tambour) faster than any human player. The other was a mechanical duck that moved its wings and neck, ate grain from a bowl, and appeared to digest the food and excrete the remains.

This really caught the public imagination. Despite Vaucanson's claims, the 'digestion' turned out to be an illusion. The grain was sucked through the mouth into a compartment inside the duck, but not digested. Instead, the excreted substance was released from a separate compartment. The mechanical ingenuity of Vaucanson's automata was extraordinary. He went on to revolutionize silk manufacture in France, one of his innovations being a precursor of the automatic Jacquard loom, controlled by a chain of cardboard strips with punched holes. At the request of Louis XV, he tried to design and make a working model of the circulation of the blood, a project that, after many years, came to nothing. In the end, the defecating duck was Vaucanson's best known achievement.

None of Vaucanson's automata has survived, but we know that eighteenth century craftsmen were extremely skilled because of three astonishingly life-like figures made by Pierre Jaquet-Droz, his son, Henri-Louis, and Jean-Frédéric Leschot. These beautiful automata, first exhibited in 1774, survived, were restored to full working order, and can be seen today in Neuchatel: a young woman playing a keyboard

A nineteenth-century print showing two of Jacques de Vaucanson's androids, the *Flute Player* and the *Pipe and Tambour Player* on their large plinths.

organ, two children, a writer and a draughtsman. As with Vaucanson's flute player, the organist's fingers move across the keyboard and depress the keys to produce the notes. Depending which set of cams is deployed, *The draughtsman* draws one of four different drawings. Three cams control the arm movement: two move the pencil on the paper left to right, and up and down, and the third lifts the pencil from the surface. The figure also blows a puff of air from its lips to remove graphite dust from the paper. *The writer* is even more complex: he moves his head and eyes, dips a quill pen in the inkpot, shakes it to remove surplus ink and then writes a message of up to forty letters. A rotating disc with steel pegs arranged around the edge controls the deployment of the cams that produce the movements required for each letter. The pegs can be rearranged to change the message. Vaucanson's automata had stood on large plinths which contained much of the mechanisms, but the Jaquet-Droz team miniaturized the components so that the whole mechanism fitted within the body of *The writer*.

The popularity of these automata encouraged many clock and watchmakers of the time to produce their own versions. Henri Maillardet, who had worked with Jaquet-Droz, exhibited a range of automata in London, including a draughtsman/writer that survived, and was restored to working order at the Franklin Institute, Philadelphia. The production of complex automata was

AUTOMATA IN THE EAST

There are very some early accounts of automata in China, such as the mechanical human figure said to have been made by Yan Shi for King Mu of the Zhou dynasty in the ninth century BCE. The account we have is from about 600 years later, and is clearly fanciful. There is also a legendary account of a *South pointing chariot* from around 2600BCE. This device consists of a horse-drawn two-wheeled chariot with a statue mounted on it, whose outstretched arm always points south, however many turns the chariot makes. It was said to be used, in pre-magnetic compass days, to navigate in the desert. A more solidly based account dates from the third century CE, when Ma Jun invented, or re-invented, a working south pointing chariot. Several more versions were reported over succeeding centuries, for some of which there are quite detailed descriptions. None survived, however, and the various reconstructions made since the mid-twentieth century are all speculative. Most use differential gearing to connect the wheels and the statue, and at least some of the historical designs may well have done the same.

Water clocks were used in China from ancient times, and by the eleventh century CE were extremely sophisticated. The water-driven astronomical clock made by Su Song was a large and complex structure with a mechanically driven armillary sphere mounted on a tower with five floors of wooden figures, rotating and playing a variety of musical instruments to mark the hours. This clock deployed an effective escapement and used a variety of mechanical components such as cams, ratchets and a continuous chain drive.

After the arrival of the first Western clocks in the sixteenth century, the Japanese also adopted clockwork, but in distinctive ways. Rather than the

A model of a *Chinese South Pointing Chariot*, following instructions published in *Meccano* Magazine, January 1957. There are now many rival suggestions for the historical mechanism.

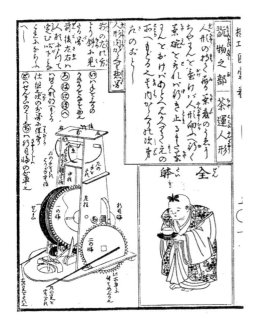

A page from *Karakuri-Zui*, the late eighteenth-century Japanese manual on clocks and automata, showing the *Chahakobi Ningyo*, or *Tea Serving Doll*, which trundles across the room, as if walking, delivers a cup of tea to a guest, and then returns with the empty cup.

European division of day plus night into twelve plus twelve equal hours, their clocks indicated day time and night time, each divided into six 'hours'. Day time and night time 'hours' would only be equal at the equinoxes. A distinctive tradition of clockwork automata or *karakuri* also developed, often with the construction entirely in wood, with cogwheels made up in segments of carefully selected hardwoods, so that the grain of the wood never ran across the teeth. Springs were constructed from baleen (whalebone). A three volume manual *Karakuri-zui*, published in 1796, contains detailed drawings of clocks and automata, including the *Chahakobi Ningyo*, or *Tea Serving Doll*.

much encouraged by a growing trade with China. The Chinese had a long tradition of increasingly complicated water clocks, with automated figures, but mechanically driven clockwork was new to them, so from the sixteenth century, the Chinese were really interested in acquiring clocks, and clockwork automata, from the West. During the eighteenth century increasingly complex and expensive machines were produced for this trade by such entrepreneurs as James Cox, whose museum of automata in London was briefly fashionable before the collection was sold by lottery in 1775, many of the pieces subsequently finding their way east.

In nineteenth century Europe the public appetite for automata continued to grow. Stage magicians such as Robert-Houdin and the Boulogne clockmaker Pierre Stevenard made some astonishing automata for their hugely popular stage performances.

New technologies and materials, and the beginnings of mass production, led to a wide range of smaller automata becoming available: some were still expensive playthings for the very rich, but other more affordable examples, including musical boxes, novelty clocks, mechanical figures and mechanical pictures, could be found in the ordinary bourgeois drawing room.

Towards the end of the century they could also be seen, on a rather larger scale, in the window displays of department stores.

Postage stamps illustrating two examples of the domestic scale automata produced in large numbers in nineteenth-century Europe.

Mechanical tin toys

Industrialization rapidly made new materials and new processes available. Clock makers developed simpler and cheaper combinations of springs, gear trains and escapements, which were also useful for mechanical toys. Tinplate, thin sheets of iron coated with tin to prevent rusting, had been in use since the sixteenth century, but, early in the nineteenth century, canning revolutionized the preservation of food, and production of tinplate increased enormously. It proved a versatile and cheap material for making toys. Recycled food containers became an important source material that could be cut and moulded into shape, and pieces soldered together, or joined with slots and tabs. To begin with, tin toys were shaped and painted by hand, but during the course of the century the availability of steel tinplate, new rolling and stamping techniques, and innovations in printing (lithography, chromolithography and offset printing) made mass production more and more possible.

By the beginning of the twentieth century tinplate toys were being made across the industrialized world. They included animals (often on wheels), carts, trains, cars and planes, but also, from many different makers such as Lehmann in Germany, and Fernand Martin in France, ingenious mechanical toys that were inspired by clockwork automata as much as by moving toys in more traditional materials.

A wind-up tin toy from the 1920s, by the German firm Schuco. The mouse's tin body is covered with felt. This is the toy that inspired Russell Hoban's classic 1967 children's story *The Mouse and his Child*.

ABOVE: **A penny toy (a later copy of an early twentieth-century design). The base is about 10cm (2in) long. Pushing the handle tips the hare forwards, and a concealed spring tips it back.**

Even simpler versions of such toys, operated with mechanisms such as levers and springs, hand cranks, inertial flywheels, or occasionally clockwork motors, could be bought on the street for one penny.

Germany remained by far the largest producer of these toys until the late 1930s.

After World War Two, production quickly recovered, with Japan becoming the leading producer.

RIGHT: ***Billy the Fisherman***, **a clockwork toy from 1951, by the British company Mettoy. The base is tinplate, but Billy, the fish and the No Fishing sign, are plastic. Driven by a clockwork motor in the base, the toy uses an ingenious escapement and will run for fifteen minutes on one winding (*see* Chapter 4 for this 'wrapping escapement').**

LEFT: **Ape with baby. One of the many small mechanical toys still produced. Most parts, including the wind-up motor, are injection-moulded plastic. This example has the added interest of a transparent body, revealing the mechanism inside.**

During the 1950s moving toys were more and more often fitted with electric motors and batteries, although wind-ups remained popular. Tin toys were still made in large numbers up until the 1960s, but increasingly plastics became the material of choice for the mass production of toys.

Earlier plastics such as Bakelite and Celluloid had been used, but as the technology developed, high volume injection moulding of

BELOW: **3D printing, the way things will go? This walking model, inspired by the amazing beach-walking machines of Theo Jansen, was designed, and 3D printed at Reading Hackspace, by Barnaby Shearer. See Chapter 6 for hackspaces.**

new, more versatile, plastics came to dominate. In the transition mechanical toys often used a mixture of materials, some tinplate, some plastic.

The nostalgic appeal of tin toys ensures that some are still made. Very large numbers of small wind-up clockwork toys are produced, principally in China. Most of the parts, including wind-up motors, apart from the steel spring and some metal shafts, are now plastic.

So, what next? The ever-quickening pace of technological development makes it difficult to predict. 3D printing, already becoming cheaper and more accessible, looks promising as a means of producing components for moving toys and automata.

AUTOMATION AND ROBOTICS

Robots were favourite post-war Japanese mechanical toys. The word, which derives from the Czech term for 'work' was first used in *R.U.R. (Rossum's Universal Robots)*, a play by Karel Čapek first produced in 1920. The robots in the play are artificial men and women made from synthetic biological materials. But the term, which had quickly caught on, was widely applied, both to automated machinery on assembly lines (the first industrial robot was deployed by General Motors in 1961) and also to *androids*, mechanical figures in which the makers, like the eighteenth century automatists, attempt to mimic human (and, by extension, animal) appearances and capabilities.

As digital computers have got smaller and faster the possibilities have expanded, but there is still a long way to go in the development of a robot that can sense its environment and move about like a human being, let alone think like one.

The artist and the machine

R.U.R was a satire on the increasingly uneasy relationship between man and machine in the industrializing world, a theme that many artists have responded to: the cartoonist William Heath Robinson (1872-1944) drew fantastical devices and absurd machines operated by serious looking men and women, to fulfil all sorts of mundane purposes. His name quickly passed into the English language to denote a ridiculously elaborate mechanical invention. In North America it is the name of the US cartoonist Rube Goldberg (1883-1970) that has passed into the language, with a very similar meaning. Heath Robinson did demonstrate a model of his pea-splitting machine on television in 1938, but most of his bizarre machinery works brilliantly only on the page. It would be impossible to build. Another British cartoonist and commercial artist, Rowland Emett (1906-1990), did produce a wonderful range of fanciful machines that actually work: After the *Far Tottering and Oystercreek Railway* for the Festival of Britain in 1951, came a series of whimsical machines, many of them commissioned to advertise products and services, such as Slumberland beds and the Post Office. They included the *Forget-Me-Not Computer* for Honeywell, which in 1966 enlivened that company's stand at the Business Equipment Exhibition by visualizing, in a fanciful mechanical contraption, elements of the computer, such as mass memory, punched card reader and random access.

In Paris, in the 1920s, the American sculptor Alexander Calder (1898-1976) made a groundbreaking series of sculptures using wire. These were a kind of 3D drawing. Calder had a degree in mechanical engineering, and many of the sculptures incorporated simple mechanisms to make them move. Above all he produced a whole circus, a collection of little mechanical figures made of wire, scraps of wood and fabric, and all sorts of recycled objects. Calder gave performances of the circus, and added components

Star Strider. A toy robot by S.H. Horikawa. Robot toys were hugely popular in Japan at the same time that industrial robots were spreading, and dreams of more intelligent and more dextrous androids were vigorously pursued.

to it, over thirty years, until it filled five suitcases. *Calder's Circus* is now preserved in the Whitney Museum, New York.

The work of the Swiss kinetic sculptor Jean Tinguely (1925–1991) displayed a decidedly ambiguous attitude towards technology. Some of his witty and ironical works, such as the painting machines, produced unpredictable outcomes; others were designed to destroy themselves in operation.

AUTOMATA TODAY

It seems to me that the Golden Age of automata is now. Different makers are influenced by different historical strands, and they come to automata by many different routes: engineering, clock making, woodworking, graphic design, fine arts, electronics and computing. This ensures a great variety of approach. Exhibitions of historical and modern automata are invariably popular, and a quick browse of the internet reveals a huge variety of images of automata.

The horizon is vast. This book will not try to travel so far. As I indicated at the beginning of this introduction, it is concerned with the design and construction of small-scale, simple mechanical devices made for fun.

Moving Folk Toys

It has to be added that throughout history, but largely unrecorded, moving folk toys were also being made all over the world in many different cultures: jumping jacks with simple string pull operation; wheeled toys to push or pull along; balancing toys; pendulum toys with swinging weights; toys operated by wind or water; toys using springs made from metal, bamboo, rubber bands or twisted string. The makers of such toys have been constantly inventive, cleverly exploiting the properties of whatever natural, re-used and recycled materials were readily available to them. They often make ingenious use of really simple mechanisms, and at their best achieve a striking and vigorous imagery.

My own route to making automata was partly through an interest in such moving folk toys. Because these toys often rely on simple mechanisms, and are made using easily available materials and a limited range of tools and techniques, I use them in this book, alongside my own automata, and those of other contemporary makers, to illustrate principles of design and construction, and the use of materials and mechanisms.

MAKING
AUTOMATA

Designing and making successful automata involves putting together a happy combination of:

Materials
Mechanisms
and Magic.

In later chapters I will explore these three elements further, but before that here are some general principles.

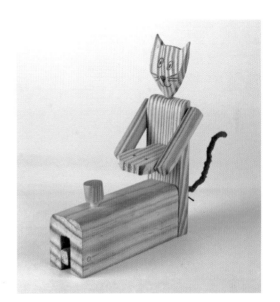

GOLDEN RULES

1 There are no rules

Automata are for fun. Making them offers a complete freedom to experiment, without many of the constraints that apply to functional structures and machinery. It is not like building a bridge, where failure may be catastrophic. You really can try anything. But of course, because you want to make real structures with real movement, *some* constraints are unavoidable: William Heath Robinson's ingenious devices work beautifully, but only on the page, where the laws of physics can be tweaked, or even suspended altogether. In the real world things can and will go wrong, or

OPPOSITE PAGE: *Muttering Bird*. **On twiddling the knob in the base, the head slightly rises and turns from side to side, which opens and closes the beak.**

Cat with Mask.

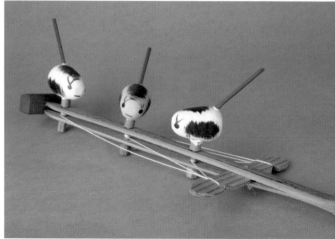

ABOVE: *Mawari Nezumi* (Spinning Mice). A traditional Japanese moving toy, showing an ingenious use of bamboo. A similar mechanism is found in a variety of Japanese folk toys.

LEFT: *Small Surprise Bird*. Simple, but effective. On lifting the base the bird turns and looks at you.

just not work at all. In which case, it's back to the drawing board (or perhaps the back of an envelope, or even the front, as Rowland Emett suggested). A good deal of trial and error may be involved in getting a new design to work properly. See that as part of the fun.

2 Keep it simple

Complicated automata are hard to make. Luckily, quite small and limited movements, made by using simple mechanisms, can be very effective. It is worth looking closely at moving folk toys, which often use such simple mechanisms in elegant and unexpected ways. The experience of making some simple automata will make it much easier to analyse and understand more complicated ones, which are, after all, made up of combinations of various simpler elements.

3 Use the properties of the materials

Materials, such as wood, bamboo, clay, wire, tinplate, paper, card, plastics and string, differ enormously in their properties of strength, durability, stiffness and flexibility. The best designs will exploit to the full the particular properties of the materials used. This is something else to look for in moving folk toys. Such toys deploy a wide range of natural materials from the particular localities where they are made, and they often show ingenious uses of recycled 'rubbish'.

4 Size matters

Just how big to make an automaton needs careful thought. One unavoidable constraint is that when you double the linear dimensions of an object the surface area will be four times as large, and the volume, and hence the weight, will be

MONKEY ON A STICK

This is a moving toy sold on the street in Indonesia in the 1990s. It could hardly be simpler. A string runs from the monkey's arms up through a loop of thin wire attached to the top of a stick, then down, through another smaller wire loop attached to the stick below the monkey, to the hand grip, which is just a scrap of card tied to the end of the string. The legs and arms are attached to rings, cut from a plastic tube, fitting loosely around the stick. When you pull the string down, the monkey goes up the stick. When you release the tension on the string, then the monkey comes down by itself, under the influence of gravity. The fact that you pull *down* to make the monkey go *up* is just enough to create an interesting gap between your movement and that of the monkey. There is just a hint of illusion that it is moving independently.

Simple as this toy is, it is worth examining closely the materials used: the stick has been split from a piece of bamboo – a readily available, cheap, strong and rigid material. The thin steel wire used to make the loops is easily bent into a shape that it will retain and bound around the bamboo stick, twisting the ends together to secure it in place. The surface of the wire is smooth – the string will run over it easily, without becoming frayed (if instead the string were run through a hole drilled through the bamboo stick, it *would* fray and break). The sharp ends of the wire could be a problem, but they are covered with what looks like small strips of metallic adhesive tape (or possibly a re-used label or packaging seal) wrapped round the stick. The string is a soft, flexible thread that runs easily through the wire loops. It is thin enough to be almost invisible, which adds to the illusion that the monkey is moving independently. The plastic rings attached to the arms and legs are also smooth, keeping the figure close to the stick, without causing enough friction to impede the movement up and down. At first glance the monkey looks like a mass-produced machine moulding, but is actually made by hand from a thin sheet of plastic foam, which

Monkey on a stick. **An Indonesian toy sold on the street.**

has been cut, bent round and butt-jointed to form the 3D body and limbs. The body is formed round a core of scrap paper. The butt joints are very neat. They were perhaps made using a solvent adhesive, but another possibility is that the joints are heat welds, made by just melting the cut edges of the foam with a heated knife blade.

Even a simple moving toy like this requires the combination of a number of different materials, each with different properties, and each doing a specific job.

eight times as large. This means that changing the size of an automaton may require changes to the proportions of the design and to the materials you use in construction.

Lion tamer. **One of the wonderful miniature automata by Laurence and Angela St Leger. Working on a reasonably small scale brings many advantages, but working at this tiny scale is really tricky.**

Ping pong players. **A Chinese clockwork tin toy. The ball flies backwards and forwards on a stiff wire (note the gap in the net). The legs of each player are firmly fixed to the base. The body, arm and bat move as one, in an arc, pushed and pulled by a rod connected to the mechanism in the base. This only crudely represents the movements involved in a real game, but, moving fast, it effectively represents an energetic rally.**

Many toys are miniature versions of artefacts, or of animals and people, and small-scale models have a very basic appeal. But there is more to it than that: by working at a 'toy' scale you gain some real advantages. Engineers seeking innovative solutions to problems often work with small-scale models, which allow them to explore a range of possibilities without the danger and expense involved in larger scale experiments. But, whereas the practical engineer has always to bear in mind the problems of scaling up from the model, the automata maker, working at a 'toy' scale, may get away with things that wouldn't work at all if they were larger and heavier. You have the pick of a very broad range of materials, techniques and mechanisms that can only be used effectively on a small scale. On the other hand it is probably best, at first, not to work on a *very* small scale, when the manipulation of materials and structures becomes fiddly.

5 Don't worry about the gaps

It is often said that the best pictures are on the radio. An artist's quick sketch may have more immediacy and vitality than the worked-up canvas. Something similar happens with automata: it's not necessary to imitate life in all the complexities of its forms and movement. A surprisingly small and simplified movement is all that is needed to make a figure seem to come alive. Nor is it necessary for an automaton to tell a long and complicated story: if you set the scene, and drop a few hints, then the onlookers will be very ready to fill in the details.

6 There is always an alternative

A design using one set of materials can always be translated into another set. There may be all sorts of reasons for doing this: the materials specified in the design could be expensive, or

Snake in a box. **Four versions in different materials. Clockwise from top left, a wooden version from India; one from Mexico with the box and the snake's head in wood, but the body of the snake in metal wire, covered with a loosely woven textile tube; one from Kenya in carved and dyed soapstone; a cut-out made from a sheet of printed card.**

not available; you might not have the skills, or the tools and equipment, necessary for working with them; you might want to change the scale, rendering the original materials less suitable; you might be seeking to improve the performance of the original design, or to change its look.

A design using one mechanism can always be adapted. Changes can always be made, but it may not be as simple as it seems at first sight. One little change will probably lead to another … and another …. Nevertheless it is a good thing to attempt such changes: you will really have to get to grips with

how the design works, and you get a greater understanding of the underlying principles.

Before moving on, it might be useful to consider how these golden rules apply to a practical example of designing and making a simple automaton.

I have taken as the starting point a piece made from driftwood called *Muttering Bird*, illustrated at the beginning of the chapter, in which a Marabou stork-like bird stands on a block of wood and moves its head slightly from side to side while opening and closing its beak. The movement is produced by twiddling a knob at the

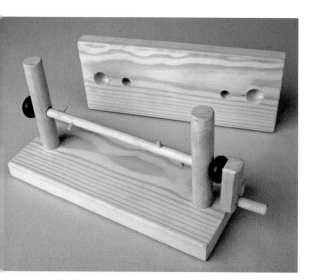

The base, partly assembled.

The bodies of the birds are made up of three layers. The two parts of the heads are hinged by gluing on a strip of soft leather.

end of a wooden shaft, which, hidden within the base, has two little pegs projecting on opposite sides. This is a camshaft, and the little pegs are cams – we look in more detail at cams, and other mechanisms, in Chapter 3. As the shaft rotates the two little pegs raise and turn, first one way then the other, a wooden disc (the cam follower), which is attached to a rigid wire run-

ning up through the leg and body of the bird to the head. The head consists of two parts hinged together. The wire is attached to the upper part, and, when the wire pushes it up, the lower part drops slightly, opening the beak.

The idea is to adapt this design in a number of ways: firstly by having an open base, with the mechanism in full view; secondly by having a *pair* of birds muttering to each other; and thirdly by changing some of the materials.

The base of the new design consists of two rectangular pieces of wood, about 160 x 65 x 12mm (6¼ x 2½ x ½in), joined with two stout dowels (a minimum of 12mm (½in) diameter, and about 75mm (3in) long) arranged along the centre line (like a two-legged table). If the dowels are seated into accurately drilled holes top and bottom, this makes a strong enough framework at this small scale. Because the two legs are on the centre line a hole drilled in each of them at the midpoint can serve as a bearing for the horizontal camshaft. It is important that the holes are accurately drilled. The dowel legs should be a reasonably tight fit, so that the base can be assembled and disassembled as the other parts are added and adjusted. Gluing the base together will be almost the last step. On a larger scale a four-sided box would make a more robust base and, again, it is best if it can be easily assembled and disassembled. The horizontal shaft is a 6mm (¼in) dowel, as in the original *Muttering Bird*, but it is elongated to carry *two* pairs of small wooden pegs (the cams). At one end is a wooden crank, as a handle to turn, in place of a knob to twiddle, and at the other end a wooden bead keeps it in place. The holes half way up the legs need to be a loose fit on this horizontal shaft, so that it turns freely.

Making the body and leg of *Muttering Bird* from driftwood involves some tricky drilling of long holes, and I wanted to see if this could be avoided. The legs of the birds are made from short lengths of bamboo, cut from a thin garden cane. Bamboo, between the nodes, is hollow, or filled with an easily removed pith. It provides a readymade strong tube. The body is made up as a sandwich of three identical pieces cut from a softwood board, about 6mm (¼in) thick (salvaged from a discarded bedframe). The middle piece is cut in three, and the central portion removed leaving a channel for the bamboo leg. Bamboo canes taper gradually along their length, so it is possible to select a bit that is just about the right diameter to fit the thickness of the wood used. The sandwich is glued together with the leg in place. The two parts of the head are made from a single layer of the same softwood board. These two parts are hinged together by gluing a small strip of soft leather at the back, opposite the beak. Before this a small slot has to be made in the lower part, through which the vertical shaft (*see* below) will pass freely. This is a bit tricky at this scale: one way is to drill a line of small holes as close together as possible and use needle files or a small craft knife to join them into a slot. After the two parts are joined with the leather hinge, a hole is drilled into the upper part of the head to take the end of the vertical shaft. The hole has to be in line with the slot in the lower part, and at a suitable angle for the beak to point down at an angle to the body. It should be a tight fit.

Drill holes of the right diameter to fit the bamboo legs in the top of the base, along the central line, so that the legs will be directly above the camshaft. They need to be placed so that the two birds will have room to mutter without clashing beaks, but far enough in from the dowel legs to allow room for the wooden discs (the cam followers) that will be attached to the bottom of the shafts that run up through the leg and body to the head. For these shafts, a *fine* split-bamboo skewer about 2mm ($^5/_{64}$in) in diameter, is used in place of the rigid wire. On a bigger piece the wire would be better, but at this scale the split-bamboo skewer is strong enough. At this stage, leave the skewers long enough to reach from the horizontal shaft, right up through

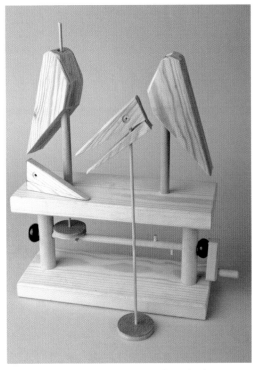

The bodies of the birds mounted on the base.

the birds with 25 to 50mm (1 to 2in) extra. Cut the discs for the cam followers from 5mm ($^3/_{16}$in) plywood. They are about 25mm (1in) diameter, and a hole is drilled in the centre of each that is a tight fit on the split-bamboo.

Measure carefully when drilling the holes in the horizontal shaft in which the little pegs are mounted – each pair must be positioned directly below the wooden disc at the bottom of the vertical shaft, with the pegs offset to each side so that they will lift and turn the wooden disc back and forth. Remember that once you drill the holes you can't adjust their position!

The pegs are cut from a 2.5mm ($^3/_{32}$in) wooden skewer. You could also use matchsticks, cocktail sticks, or pieces of the split-bamboo skewer, but your drill bit must match the diameter of the material. Another possibility is to replace the pegs with little egg-shaped cams cut from a thin piece of wood, such as the plywood used for the wooden discs. Drill a hole in each to match the diameter of the camshaft. It is best to drill these holes *before* you cut the cams to shape. They can then be slid onto the shaft, and you can adjust their position before finally fixing in place with a touch of glue.

In the final assembly, leave the heads until last. When the other parts are glued in position, having checked that the camshaft rotates freely, and that the vertical shafts are turning to and fro,

position the heads and glue them to the vertical shafts. The shafts must first be shortened so that at their lowest position, when they are not lifted by the cams, the heads just rest on the shoulders, with the beaks closed.

In a second version of the new design the body and head are constructed as in the original *Muttering Bird*, replacing the driftwood with hardwood offcuts. The head is hinged by cutting

Muttering Birds completed.

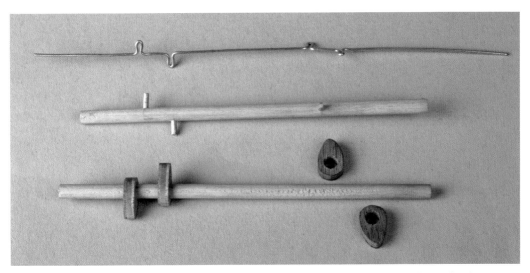

Three alternatives for the camshaft. Dowel, with holes drilled to take smaller dowel pegs for the cams (centre); dowel, with egg-shaped wooden cams with holes just large enough to slide onto the shaft (below); a single piece of galvanized steel wire bent to form the shaft and the cams (above). In each case one pair of cams is set at right angles to the other so that the movement of the birds will alternate.

a slot in the top part, and gluing in a thin piece of wood, which projects below into a longer, and *slightly* wider, slot cut in the lower part of the head. A brass pin fixed transversely through the back of the lower part runs through a hole in the projection, holding the two parts together, but allowing the beak to open. Glue metal beads into shallow holes for the eyes. The body is a solid piece, cut and sanded to shape, with a hole drilled right through it to take the leg, which is a bamboo tube, as in the first version.

The vertical shaft, again as in the original *Muttering Bird*, is a rigid piece of wire (in this case cut from an old bicycle spoke), but in place of the horizontal wooden camshaft use a single piece of galvanized steel wire. It is bent in four places into a tight 'U' to form the four cams. The scale is important here, because if the wire is too thick it will be really difficult to bend neatly and accurately, but if it is too thin then the camshaft will not be rigid enough. As with the wooden shaft with little pegs glued in as cams, accurate measurement is important. It is quite tricky to end up with a straight shaft and the two pairs

of projecting cams exactly positioned in relation to the vertical shafts. Make the bends for the cams first, check the measurements, and feed the shaft into the holes in the dowel legs, before

The birds for the second version. Each body is shaped from solid wood, and a hole drilled through for the leg. The head is hinged with a tongue set into the upper part, and projecting into a slot cut in the lower part.

Muttering Birds.
The second version
completed.

you reassemble the base. After this a bead can be fed onto each end and the wire bent into a crank handle at one end, and into a small loop to secure the bead at the other.

Looking at this exercise in respect of the six golden rules, there are some points to note:

1 Doubling up the birds is an experiment. I like the resulting look of *Muttering Birds*, but how about the way it moves? Operating it by turning a crank handle has a different feel to twiddling a knob. Possibly because each bird tends to mirror the other, the movement seems a little more regular, more stilted. Turning the handle very slowly seems to give a better result, and so does turning it very fast, which gives a manic effect that you could not get by twiddling the knob on *Muttering Bird*.

2 As with *Muttering Bird* the mechanism is quite simple, but gives good value in terms of movement. The back and forth movement from the pair of cams not only turns the head, but also, as it lifts, allows the beak to open

simply through the action of gravity. One problem in using a pair of offset cams to give a back and forth rotation to a vertical shaft is that the shaft must be completely free to move up and, more particularly to fall back *down* under its own weight, but at the same time must be held vertically upright. In this design the leg and the hole through the body of the bird together form a *long* vertical tube which can be made a really loose fit on the shaft, but still hold it upright. Another problem is that a slight difference in the length of the cams, or a slight difference in the offset from the centre will result in the shaft gradually rotating round, one way or the other, through 360 degrees. A solution often used, if this movement is not required, is to attach a projection to the cam follower, which comes up against stops on the base frame. This will prevent the full rotation. In this design, however, this is not necessary, because the shape of the birds' shoulders, the off-centre positioning of the heads, and the angle at which they slope, tend always to return the heads towards the front of the bodies.

3   The distinctive properties of various materials are exploited in the two versions of *Muttering Birds*: e.g. the bamboo tubes; the strength and rigidity of the split-bamboo skewers; the capacity of the galvanized steel wire to be bent into a shape that it will then retain.

4   *Muttering Birds* is slightly smaller than *Muttering Bird*, allowing the use of some features that might not work so well at the larger scale. For example: the simple two-legged structure for the base, and substituting a split-bamboo skewer for the bicycle-spoke wire for the vertical shaft.

5   The birds have no wings, one rather thick leg, no neck, a large strange head and beak, and a very upright stance, dictated by the fact that the head must be positioned directly above the leg for the mechanism to work. Nevertheless, once the head moves a little from side to side, and the beak opens a bit, the overall impression is surprisingly life-like. It reads quite convincingly as a stork or heron-like bird of a kind that often does stand rather upright, possibly on one leg, and tucks its neck up so that the head rests on its shoulders. In my experience with *Muttering Bird* I have found that aficionados of Marabou storks will recognize it as one, whereas lovers of herons will claim it as theirs.

6   A number of materials have been changed, including the softwood board, or the hardwood offcuts, for the driftwood of the original design, and the galvanized wire camshaft in place of a wooden one.

Even minor changes to the mechanisms have consequences.

The movement is powered by turning a crank handle, instead of twiddling a knob. The design must allow a sufficient depth to the base to accommodate the chosen length of the crank handle when it turns through 360 degrees. In the second version of *Muttering Birds* the head is hinged with a tongue projecting from the top part into a slot in the lower part. That slot can be cut in from the back, being made long enough to accommodate the tongue of the hinge, and also leave a gap for the upright shaft to pass through in front of the hinge. The lower part has to be slightly rounded off at the back to allow the hinge to open, pivoting on the brass pin which passes through the tongue. In the first version, on the other hand, the two parts of the head are both left square at the back, where they pivot on the leather hinge. A slot in the lower part is still required, allowing space for the vertical shaft to pass through, but it is better if it does *not* extend to the back, where the leather is glued on, so a set-in mortise has to be cut.

MATERIALS

Choosing the materials for simple small-scale automata can be largely a matter of what you have readily available, together with your tools and skills. Even a very simple design is likely to combine a number of different materials. In *Balancing Bird*, for example, the base is a piece of driftwood, which I like using, firstly because collecting it involves business trips to the seaside. Secondly I like it because its previous use often still haunts it, and its time at sea builds its character. In this case the wood is teak, burnt at the end, and it came from the flooring of the West Pier at Brighton, which was destroyed by fire. It was washed up much later, miles away, on Chesil Beach. The bird is also driftwood, with the wings and tail made of guinea fowl feathers, imported from Africa. The balanced lever, on which the bird rotates, is stainless steel welding wire, and the counter-weight is a bicycle spoke nipple. The little metal cup at the top of the base, on which the lever rotates, is made of a brassed steel paper fastener.

WOOD

A majority of the automata and moving toys illustrated in this book are, at least in part, made of wood. Wood is widely available both as new timber and for re-use and recycling. It is relatively easy to cut, drill, and fix with limited tools, producing simple, solid structures. But its potential is virtually unlimited. With advanced skills and more sophisticated tools, you can work it into beautiful, complex and varied structures.

Wanda Sowry's *Dia de los Muertos* is an all-wood automaton, with a simple, but solid, oak

OPPOSITE PAGE: **Balancing Bird. The base is teak from the West Pier at Brighton, destroyed by fire.**

Wanda Sowry's *Dia de las Muertos* uses different woods to great visual effect.

Talking Birds **exploits the weathered appearance of driftwood. The orange handle is shown depressed here, raising the other end of the lever and letting the neck of the bird on the right fall forward.**

framework. A central shaft, turned with a crank handle, bears six cams which control the movements of a skeleton band.

In this piece, differently coloured woods have been used very effectively. No paint was needed, but wood does also lend itself well to colouring with paint, and with dyes. Many different styles of finish will be found in the illustrations throughout the book.

Talking Birds is made from driftwood. The birds stand on an oak stave from a barrel. The varied colours of the driftwood are contrasted with the bright-coloured plastic rope and touches of paint on the birds' crests and eyes, and on the handles at the ends of the operating levers.

OTHER PLANT MATERIAL

All sorts of plant material other than wood are used in moving folk toys, and can be very useful for making automata. For example:

Bamboo is a cheap and strong material. Being hollow between the nodes it is an excellent source of small tubes, and skewers made from split bamboo are a good substitute for small wooden dowels. Wherever bamboos grow they are used for an extraordinary range of products. A Japanese folk toy depicts the hero Benkei in a coracle-like boat. The bamboo mechanism neatly exploits the properties of the material. A strip of bamboo is split in two, back to a node, which remains intact, making a framework to hold the operating lever and the shaft supporting the coracle.

An elegant bamboo toy from Japan. A similar framework is seen in the spinning mice toy, illustrated in Chapter 1.

Weaving structures with plant materials is an ancient technique. There is considerable technical skill in a Malawian push-along toy helicopter. The structure and the moving parts are all woven from grass, with some rigid shafts of split bamboo. The front axle bears a single woven wheel, which also acts as a friction drive cam, moving a woven disc at the bottom of a vertical shaft which has the rotor blades on top. Pushing the wheels along the ground also turns the rotor.

Rodney Peppé's *Singing Nut* is a clothes-peg toy that makes witty use of a walnut shell. The shells are hinged at the back with a strip of leather. The lower half-shell is mounted on the upper arm of the peg, and a string is attached to the top of the upper half-shell, which passes through a hole in the upper arm to be fixed to the lower arm of the peg. When the peg is squeezed, the 'mouth' opens revealing a strip of musical score.

Helicopter toy from Malawi. Although made mostly of grass it has a friction drive mechanism to turn the rotors as it is pushed along.

Rodney Peppé's *Singing Nut*. A delightful combination of a clothes-peg and a walnut shell.

In India, various moving toys are made from shola, or sola pith. This is an easily cut, very white, light, even-textured pith, found in the stems of a legume, *Aeschynomene aspera*. An animal, made in Assam by Manindra Malakar, is a very effective example of a 'nodding head' toy. The hollow body, the four legs, the head and the tail are all made of the lightweight pith. The limbs, head and tail are suspended by a thread, and inside the body each has a small blob of clay on a sliver of split-bamboo as a counterweight.

CLAY

Apart from its use as a weight, as in the sola pith animal, ceramic material is not very much used in moving toys or automata, although there is no reason why small, simple mechanisms should not use ceramic components. In a traditional toy drum cart from Bangladesh, the wheels and the drum are clay, mounted on a bamboo framework held together with string. The axle is a camshaft, with cams made from thin strips of bamboo slotted through the shaft. The cam followers are beaters that strike the paper skin of the drum as the cart is pulled along. *See* Chapter 3 for a discussion of cams and followers.

METAL

Metals are generally less easily worked than wood and other plant materials, but wires, rods and tubes, and thin sheets, such as tinplate, are versatile components of moving toys and automata.

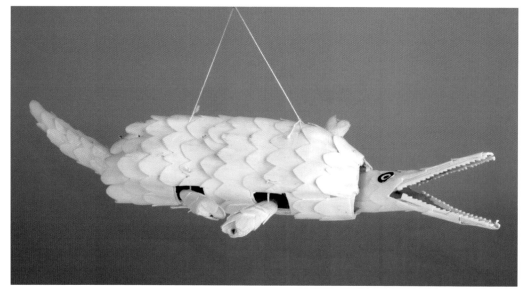

A toy animal from Assam, made of light-weight sola pith, that nods not just its head, but its tail and all four limbs.

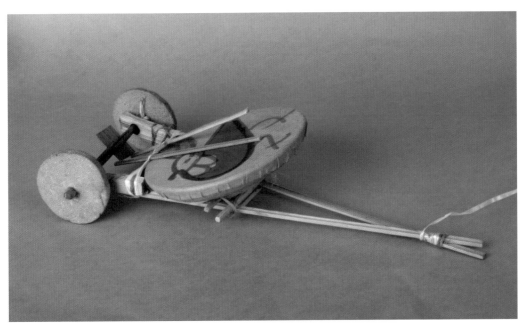

A traditional drum cart from Bangladesh. The drum and the wheels are made of clay.

Tinplate was the dominant material for mass-produced mechanical toys from the late nineteenth century until the 1960s. Very often the component parts were assembled with a slot and tab technique. A series of tabs along the edges of the parts are fed through small slots in the adjacent parts, and bent over to hold the pieces together. A Malian push-along toy helicopter is an interesting example of this technique, made entirely by hand. The mechanism is very much the same as that used in the Malawian grass helicopter. The front axle, made of wire, has a tin disc. On this rests another disc, at the bottom of a vertical wire shaft that carries the rotors. When the wheels are pushed over the ground this friction drive (*see* Chapter 3) transmits motion to the rotors, which spin round. The tab and slot construction is not the only interesting technique used in this quite rough looking toy. The tinplate from which the rotor blades and discs are cut is thin and inelastic, making it very difficult to attach firmly enough to the wire shafts without glue or solder, which will have been unavailable to the maker. The solution used is to sandwich

each of the tinplate components between two little squares cut from the walls of a plastic container of some sort. This plastic is quite thick and fairly rigid, but when pierced with a small hole and pushed on to the wire it is elastic enough to grip firmly and stay put. The same technique has

A tinplate helicopter from Mali, assembled without the use of glue or solder.

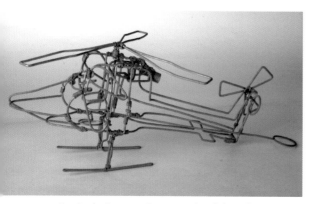

A wire helicopter. An example of the African technique of making a 3D drawing in wire.

been used to secure the wheels, made from the soles of old flip-flops.

There is a long tradition in many parts of Africa of building toys and models from locally available plant materials, using not only wood, but fruits, seeds, palm leaf petioles and herbaceous stems from crops such as sorghum and millet. The latter, for instance, were built into quite complicated structures, such as model vehicles, using lengths of the whole stem pinned together with slivers of the hard outer layer. During the course of the twentieth century the same ingenuity was applied to newly available materials,

Ian Mckay's *Pecking Order.* A simple, but elegant combination of materials.

such as fencing wire. Other materials might be added, but often the wire was used alone, sometimes with several different thicknesses combined in one model. The technique is to make a sort of 3D drawing in wire, reminiscent of Alexander Calder's sculptures made in Paris in the 1920s. The subject is often a vehicle of some sort – a car, a truck or a helicopter – with moving parts, such as wheels, axles and rotors.

The bent-wire technology has also been used to great effect in combination with painted wood, tinplate and other materials, as in W*omen Pounding Millet*, illustrated in Chapter 3.

Ian McKay's *Pecking Order* uses a similar mix of materials. The elegant wire work of the mechanism complements the carved and painted limewood of the chickens, and the half section of copper pipe as a feeding trough.

SOME OTHER MATERIALS

Paper and card

Paper and card provide an extremely versatile medium for automata. The first project in Chapter 5 is a relatively simple cut-out, and I shall discuss some of the issues in designing cut-outs there. A wide range of cut-out automata is available as printed sheets, or to download from the internet (*see* Chapter 6). *Feeding Time at the Zoo* is one of Tim Bullock's hilarious designs for *cool4cats*. The crank handle turns a belt carrying an endless supply of school children from the bus to the jaws of the crocodile, which open and close to consume each one.

Feeding Time at the Zoo, one of Tim Bullock's witty and effective designs for cool4cats.

Fabrics

Fabrics, combined with wood and wire work for the mechanisms, can add an unexpected element, as in Janine Partington's *Rabbit in a Hat*, where the hat and the rabbit are both made of felt.

Plastics

Injection moulded plastics have replaced printed and stamped tinplate as the dominant material for mechanical toys. Now developments in 3D printing look set to bring more accessible ways of producing small mechanical devices, in plastics, and other materials.

Meanwhile plastics have replaced wood and metal for a wide range of everyday objects.

Plastic detritus in the oceans poses a growing problem, and beachcombers like myself find that, for example brush heads, once wooden, are almost all plastic now. In *The Rong Sisters*, a large piece with two aeroplanes flying round a central pole, I used a few of these plastic objects from the beach to make the aeroplanes.

Ready-made parts

The automata maker can find a use for all sorts of stuff, most of it made for other purposes, such as the bicycle-spoke nipple used as a counterweight in *Balancing Bird*, the copper pipe in *Pecking Order*, the tin can in *Women Pounding Millet*, and umbrella parts used for the oars in *Rower*, illustrated in Chapter 3.

Rabbit in a Hat. Janine Partington's use of sewn felt for the hat and the rabbit is attractive and unusual.

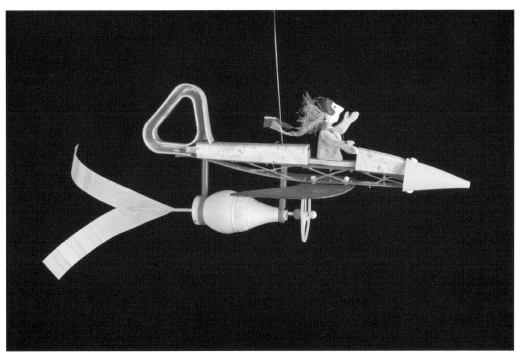

Audrey Rong in her plastic plane. A detail from *The Rong Sisters* showing the use of plastic rubbish, joined together with plastic pop-rivets.

Fixing things

Automata have moving parts, and present two different problems: making joints that are fixed and making joints that move, often in a particular constrained way.

For the fixed joints that do not move, it is useful to be able to take things apart during construction. Wooden joints can be dowelled or pinned and glued at the final stage, or parts may be screwed together. There is a bewildering range of adhesives available, many of them rather toxic. My personal choice is white PVA wood glue for wood, paper and card, and a two-part epoxy resin where PVA is unsuitable, such as wood to metal, or metal to metal joints.

Where holes are drilled in a wooden frame, or through a guide piece, to act as bearings for a wooden shaft, the holes need to be big enough to allow the shaft to move freely. Remember that the wood will shrink and expand slightly, according to temperature and humidity. On the other hand, if the hole is too big the shaft may rattle around, or be pushed sideways and jam. Applying a dry lubricant, such as graphite powder (sold for freeing car locks) may help. Rubbing the point of a graphite pencil on the wooden shaft, or in the bearing, is almost as good. Talcum powder is a less messy alternative. However, it saves an awful lot of bother if you can make holes the right size. This applies both to loose holes made as bearings for moving shafts, and to holes made to be a snug fit, for gluing in dowels or to mount components such as cams on a shaft. It is well worth investing in a set of drill bits with as small as possible jumps between sizes. Another useful tactic is to line a wooden bearing with a plastic tube. PTFE (polytetrafluoroethylene) tubing works exceptionally well, but almost any plastic tube, such as a drinking straw, or a ball point pen casing may be useful.

MECHANISMS

A mechanism is a group of interlinked moving parts that perform a function.

THE FIVE SIMPLE MACHINES, OR IS IT SIX ... OR TWO?

The ancient Greeks classified the basic elements of useful mechanical devices as five 'simple machines'. These were the INCLINED PLANE, the WEDGE, the SCREW, the LEVER and the WHEEL. In the Renaissance the PULLEY was added to these 'mighty five'. The inclined plane, the wedge and the screw are closely related, as are the lever, the wheel and the pulley – thus you end up with only two distinct elements, by combination of which more complex machines arise. This led Franz Reuleaux, in the nineteenth century, to suggest that it would be better to look at and classify JOINTS, the connections that provide movement, as the primary elements of a machine. A *kinematic chain* consists of links, rigid in themselves, but connected by joints allowing different movements. Getting the joints between parts to move in just the appropriate way is often one of the principal problems to be solved in making automata.

OPPOSITE PAGE: *Posh Bird*. **A combination of three levers.**

USEFUL MECHANISMS

Automata are machines in which power is applied through a mechanism, or a combination of mechanisms, to achieve the required movement. All sorts of different mechanisms may be used, such as levers and linkages, cranks, cams, wheels, gears, pulleys, springs, ratchets and pawls.

Levers and linkages

A lever is a rigid bar pivoted at some point along its length (the fulcrum). Using a lever, a smaller effort (the force applied) can be used to move a larger load, if that load is closer to the fulcrum and the effort is applied further away. Archimedes, speaking of levers, is supposed to have said 'give me a place to stand on and I will move the Earth'. Of course the mechanical advantage of using a lever in this way is gained at the expense of applying the smaller effort through a greater distance. At the small scale of moving toys and automata the reverse effect is just as useful, where a larger effort moves a smaller load through a greater distance. Levers are often divided into three classes, according to the arrangement of fulcrum, load and effort. Class One levers have the fulcrum between the load and the effort as in a see-saw. In Class Two levers the fulcrum is at one end, the effort at the other, and the load in between, as in nutcrackers, or a wheelbarrow. In Class Three levers the fulcrum is at one end, the load at the other end, and the effort in between. The lever that is pushed down to operate *Cat with Mask*, illustrated in Chapter 1, is a Class Three lever.

Posh Bird has a simple mechanism consisting of three linked levers. A lever runs horizontally in a slot cut through the base, pivoted close to the end which forms the handle. The other end has a stone weight hanging from it. When the handle is depressed the weight is raised, and a string, running from this lever up one of the bird's legs (which is a hollow brass tube), and through the body, releases the tension on the lower end of the neck. This is another lever, which the weight of the

Peter Markey's *Walking Dog* produces a rather complicated-looking movement from a single crank.

head now causes to fall forward. As it does so the beak opens, because the upper part of the head is a third lever, pivoted on a hinge at the back and connected with a thread to the top of the body. As the neck tips forward, this thread, attached just behind the hinge point, is tightened, and pulls the beak wide open. When the handle is released the stone falls back down, the neck of the bird is pulled up again, and the beak closes.

Cranks

A crank is a type of lever consisting of an arm sticking out from a shaft, which communicates movement to or from that shaft. If the arm is bent back again, parallel to the shaft, the extension forms a crank handle which can be used to turn the shaft, as in many of the automata illustrated in this book. The longer the crank arm, the more the mechanical advantage, and the easier the handle is to turn, but also of course the further the handle has to be moved for each rotation. If a connecting rod is pivoted on the extension it will move backwards and forwards when the shaft rotates, the total distance moved being twice the length of the crank arm. This distance is known as the *throw* of the crank.

In Peter Markey's *Walking Dog* the shaft mounted on the right has a crank handle to turn at one end, and a shorter crank at the other end, with a connecting rod attached, which is linked to the body of the dog, sliding it back and forth on a stationary inner frame. This reciprocal movement operates the six ingeniously articulated levers that make up the head, tail and four legs of the dog, producing the effect of a vigorous walk.

In Aquio Nishida's *Hound Dog* the body of the dog is mounted rigidly on a connecting rod moved up and down by a crankshaft mounted in the base. The movement of the rod is guided by passing through a hole in the top of the base. As the crank turns on a circular path it passes directly below the guide hole only at the highest

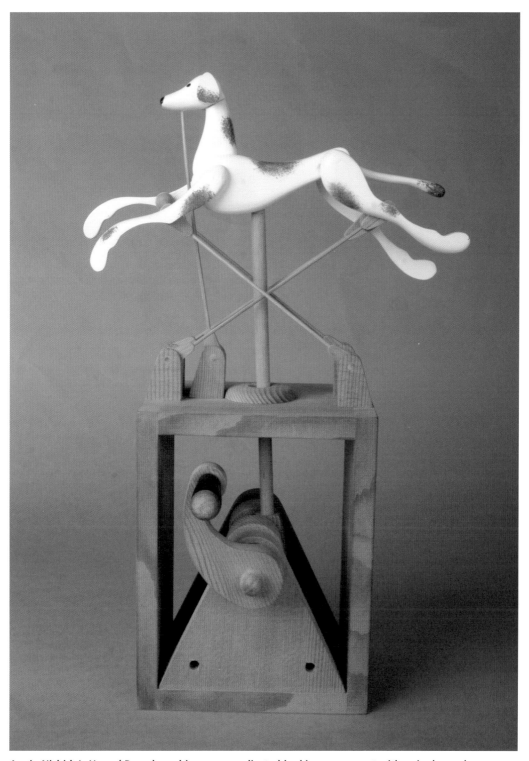

Aquio Nishida's *Hound Dog* also achieves a complicated-looking movement with a single crank.

In *The Rower* the path taken by the connecting rod, and the boat attached to it, is constrained by a link, making it tip just enough to rock the rower back and forth.

Detail of the wire helicopter illustrated in Chapter 2, showing the crank.

and lowest points. In between it travels to one side or the other. This causes the connecting rod, and hence the body of the dog which is attached to it, to move backwards and forwards, as well as up and down, when the handle is turned. The distance from the crankshaft to this guide is critical to the design. Changing that distance does not alter the up and down movement of the connecting rod. The extent of that is the throw of the crank, determined by the length of the crank arms. But if the guide were moved closer to the crank, the backward and forward movement would be greater, and if it were moved further away it would be less. Placed as it is, the body describes an elegant loop. The limbs and head are loosely articulated to the body, and joined with slender links to the base, so that they move in relation to the body in a convincing running motion.

In *Rower* the boat is mounted on a connecting rod joined to a crank consisting of a wooden disc with an offset pin. The movement of the connecting rod is constrained here not by passing through a guide hole, but by a rigid link attached to a post fixed to the base, all the joints being free to rotate.

The effect of this linkage is a movement similar to that of the body of *Hound Dog*. As in that piece, the extent of movement up and down is fixed by the throw of the crank, but here the amount of backwards and forwards movement depends on the length of the restraining link, and its point of attachment to the connecting rod. The length of the link, and the attachment point, are chosen so that the boat rocks just enough to tip the loosely mounted rower back and forth. His arms are rigidly attached to the body, while the oars are loosely attached to the hands, and free to slide in the rowlocks, so that as the body tips he appears to row.

In these examples a crank is turned to move a connecting rod. In the wire helicopter illustrated in Chapter 2 a connecting rod is used to turn a crank, which is what happens in an internal combustion engine, where the connecting rods, driven by the pistons, turn the crankshaft. In the wire helicopter the vertical shaft from the rotors is bent into a crank. A connecting rod linked to this crank sticks out at the back, and can be moved backwards and forwards by hand, thus spinning the rotors.

Wheels

In *Rower* the crank is a round disc, fixed at the centre to a shaft, and with a connecting rod mounted on a pin offset from the centre. In a push-along toy bird from Portugal a pair of similar round discs act both as cranks and as wheels. Each wheel has a wire link, similarly offset from the centre, and attached to a wire loop on the underside of the wing. When the toy is pushed along the ground the wheels turn, moving the wire links up and down. The rotary motion of the

Wooden push-along from Portugal. The wheels also act as cranks.

wheels, acting as cranks, is converted to reciprocating movement of the links. This causes the wings to flap. The wings are hinged at the body, and move together since the wire links from the two wheels are aligned.

Women Pounding Millet. **A Zimbabwean push-along toy with a wire crankshaft and linkages.**

A bell crank.

In an African push-along toy showing women pounding millet, two wooden wheels are joined by a wire axle which is also a crankshaft. The cranks are linked with wire connecting rods to wire extensions from the arms. These extensions are at an angle to the shaft joining the arms, so that the back and forth movement of the connecting rods is converted to an up and down movement of the arms.

These bent levers, very useful for changing the direction of a movement, are known as *bell cranks* because of their old use in linking wires around corners, from a bell-pull to a bell in another part of the house (for a front door, or to summon servants).

Cams

A cam consists of projections from a revolving shaft that convert the rotation into the linear movement of a follower. The shape of a cam can be very simple, such as the little wooden pegs in *Muttering Birds* (*see* Chapter 1). They can also be very complex, and

combined in numbers. Simple or complex, cams represent a kind of physical memory of a sequence of movements. In a mechanical barrel organ there might be thousands of little pins and projections arranged in rows around the barrel, which 'remember' a series of tunes. Similarly, several series of cams with complex profiles are needed in a clockwork android such as the Jaquet-Droz *Writer* described in the introduction.

In *Rabbit in a Hat*, illustrated in Chapter 2, there is a single, very simple, cam, consisting of a circular wooden disc mounted eccentrically on the shaft, which is turned with a crank handle. The follower is a vertical rod that slides up and down in a hole through the top of the framework. This guide hole must be wide enough to allow the follower to move freely up and down as the cam rotates, but if it is too loose, the cam will tend to push the follower sideways so that it jams. The longer the guide is, and the closer it is to the cam, the better it is likely to work. The follower has a wooden disc at the bottom, which makes a broader surface to rest on the cam. This will help if the follower is not placed absolutely centrally above the cam, although in that case the follower may tend to rotate as well as go up and down. In the case of a rabbit popping out of a hat, that does not really matter. In Kasuaki Harada's *Teddy Bear* the arrangement is similar. Here the bear's body is fixed to the base on a copper tube. That tube, together with the hole through the body, makes a long guide for the follower, which is attached to the head. In this case the moving head is linked to the static body by the jointed arms, which prevent rotation, or very nearly do so. As the head goes up and down there is just a hint of movement sideways, constrained by the arms, giving a tweak to the head which adds a certain liveliness to the rather macabre effect.

If the cam is moved to one side of the disc at the bottom of the follower, then the follower will rotate, an arrangement known as a friction drive. Where the cam is a circle mounted centrally onto the camshaft then the follower will simply rotate, without

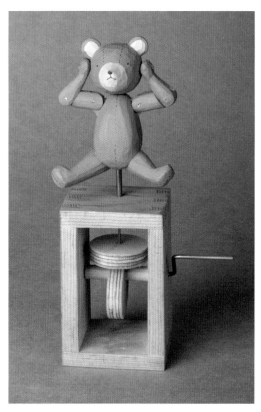
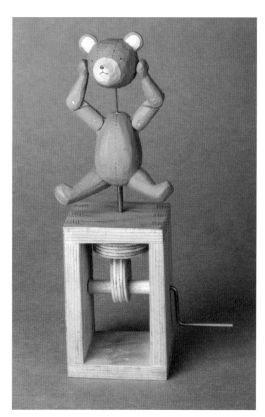

Kasuaki Harada's *Teddy Bear*, operated with a single eccentric cam.

going up and down. This mechanism is used to turn the rotors in both the tinplate and the woven grass helicopter illustrated in Chapter 2. In Peter Lennertz's *Fishing Cat* there is a pair of eccentric circular cams set either side of the disc at the bottom of the follower. The upper part of the cat is alternately lifted and turned right and then left. A hole through the fixed lower part of the body provides the guide, and a dowel projecting from the follower disc at the back catches against the framework to limit the rotation. The fishing line with the jiggling fish is also attached to the framework.

In *The Motley Crew*, illustrated at the beginning of the book, there is a camshaft with seven cams, each with a follower that takes the form of a lever, attached with a string or wire link to

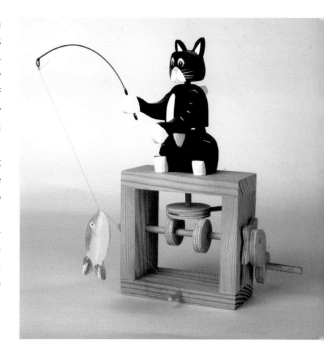

RIGHT: **Peter Lennertz's *Fishing Cat*, operated with two offset cams.**

Detail of *The Motley Crew* (illustrated at the beginning of the book), showing the cam shaft and bevel gears.

the part to be moved. Cam followers in this form often need something, such as a spring, to hold them firmly in contact with the cams. In this case two of them are below their cams, and are pulled into contact simply by the weight of the loads. The other five run above the cams and are held firmly in contact by stone weights that hang from them.

Gears

Toothed gears are another way to transmit power. The movements of eight figures are controlled by the seven cams in *The Motley Crew*, and if the camshaft were directly turned by hand things would happen too quickly. The load from eight moving figures adds up too, so turning the shaft directly might be quite hard. The solution is to use a separate driveshaft connected to the camshaft through a set of bevel gears. These were bought at a car boot sale, because they looked as if they

would come in handy, and are, I believe, part of a marine steering mechanism. The small cogwheel has to be turned about four times for one revolution of the larger cogwheel. The mechanical advantage makes the handle easier to turn, and slows thing down quite nicely as well.

Pulleys

A pulley is a wheel that turns about its axis, carrying a belt or cord, and can be used to raise a load, transmit power, or change the direction of a pull. In the tinplate see-saw toy a small pulley, mounted on a shaft with a crank handle, is connected, with a rubber band forming the belt, to a larger pulley on a second shaft. This second shaft has a crank that is linked by a connecting rod to the see-saw. The small pulley is about two and a half times smaller in diameter than the larger pulley, so the crank handle has to be turned about two and a half times to rock the

RIGHT: A small tinplate see-saw with a belt and pulleys to transfer the drive.

BELOW: *Falling Owl*, showing the use of pulleys to change the direction of a movement, and to rotate a shaft.

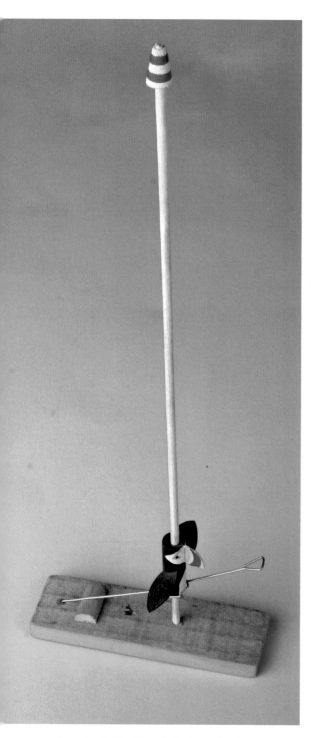

see-saw once. On this small scale the mechanical advantage thus gained is not important, but the slowing down of the action is useful.

Falling Owl has a string attached to the top of its head which passes over a pulley in a block suspended above, and down again through a hole in the body of the owl. Here it passes once around a small pulley mounted on a shaft that links the two wings, and out at the bottom, where it is attached to a piece of wood acting as a handle. When that handle is pulled down the owl is pulled up towards the pulley block at the top, and the wings spin. On letting go of the handle the owl falls under its own weight, its descent moderated by the counterweight of the handle, and by the governing effect of the spinning wings. The upper pulley both acts as a support, and changes the direction of the force applied by pulling the string down, or by the falling weight of the owl. The pulley inside the body transmits movement from the string to the wing shaft.

Springs

As mentioned above, a spring is often used to hold a cam follower in contact with the cam. A rubber band is used in this way in the second project in Chapter 5. In *Jumping Puffin* a wire spring provides the motive force to flick the hollow body of the puffin up the stick.

Many moving toys that are operated by hand with a squeeze or a push use a spring for the return movement. The pecking chickens and the boxers illustrated in Chapter 4 are examples of this. On a small scale a clothes-peg offers a convenient little platform, consisting of two levers hinged together with a spring. *Singing Nut*, illustrated in Chapter 2, uses one. *Black Bird* is mounted on the upper arm of a peg. The wings are fixed to a shaft through the body, which is free to pivot. A wire link joins the lower arm of the peg with a wing, in front of the pivot point. When the peg is squeezed the wire link is pulled down, the back of

Jumping Puffin. The steel wire spring is depressed, and then released suddenly by sliding the finger off the end, to make the puffin jump.

Black Bird stands on a clothes-peg, which makes a convenient spring return device.

the wings is raised, and the movement of a little cam on the wing shaft inside the body lets the beak fall open. On releasing the peg, the spring closes, the wings drop back and the beak shuts.

Ratchets and Pawls

In *A History of Aviation: The Age of Driftwood*, illustrated in Chapter 4, a rotating shaft, with arms supporting two aeroplanes, is turned by pulling a cord that passes around a pulley. The pulley is free to rotate on the shaft, and the drive is transmitted through a ratchet and pawl. The ratchet consists of two metal pins projecting from the pulley, just above where the cord runs. The pawl, fixed to the main shaft, is a little lever that hangs down so that it can engage with the pins. It is free to swing up to the right, but there is a metal pin, just below the pivot point, on the left which stops it swinging to the left. When the cord around the pulley is pulled to the left the pawl locks against one or other of the pins, and the main shaft rotates. When the cord is released it is pulled back towards the right by a counterweight, but the pawl is free to swing up each time it meets one of the metal pins of the ratchet, so the shaft continues to rotate clockwise, almost unimpeded. This has very much the same effect as a freewheel on a bicycle. A ratchet may also be used as a locking mechanism to prevent a shaft turning the wrong way, or, if the

pawl is made to move backwards and forwards, for an intermittent drive.

Another intriguing intermittent rotary drive is the Geneva mechanism. This can be seen in the illustration of *Man and Marble* in Chapter 6. A dowel on the drive shaft connects with a slot in the Maltese cross on the driven shaft, and four revolutions of the drive shaft give four quarter revolutions of the driven shaft, separated by four periods at rest.

Detail of *A History of Aviation, the Age of Driftwood* (see Chapter 4), showing the ratchet device in the drive.

MAGIC

From the beginning, automata were used to demonstrate priestly and magical powers and generally to mystify and astound. There may then have been many in the audience ready to ascribe supernatural powers to the machines and their makers, but one suspects that even then automata were objects of amusement as much as of mystery and awe. But powerful and enigmatic objects. Machines as animate beings, animate beings as machines, these are ideas that have always evoked deep and uneasy responses. Automata took their place in the caliphs' courts, in pre-Reformation churches and in the gardens and palaces of Renaissance princes. In the eighteenth century clockwork androids and automata fascinated everyone, including European kings, Chinese emperors and Japanese Shoguns. Clock and watchmaking skills developed so that complex mechanisms could be made smaller and smaller, firstly being used in a huge variety of more affordable domestic scale automata, and eventually becoming cheap enough to use in mass produced mechanical toys. In the nineteenth century conjurors such as Robert-Houdin used automata to create the *illusion* of magic, in the sense of supernatural power. But the best automata always displayed magic in another less literal sense: an unanalysable charm, which neither a growing ambivalence towards the machine, nor modern developments in computer science and robotics have diminished.

Automata have retained their magic. Not, literally, a supernatural power, but a strong fascination that derives from the combination of many different factors.

An effective automaton is a whole that is more than the sum of its parts. It is difficult to pin down the precise qualities that contribute to a design that produces this magic result. Here are some observations:

YOU DON'T HAVE TO HIDE THE MECHANISM

In the clockwork marvels of the eighteenth and nineteenth centuries the mechanisms were largely hidden. But, mysteriously, the magic works even when the mechanism producing the movement is in full view. Indeed, a visible mechanism often adds considerably to the total effect. The movement still, to some degree, makes a mechanical figure seem to come alive. The ambivalence between the mechanical and the animate remains strong, and in addition the aesthetics of the machinery come into play, whether spare and elegant, or absurd and unexpected.

OPPOSITE PAGE: *Thinks.*

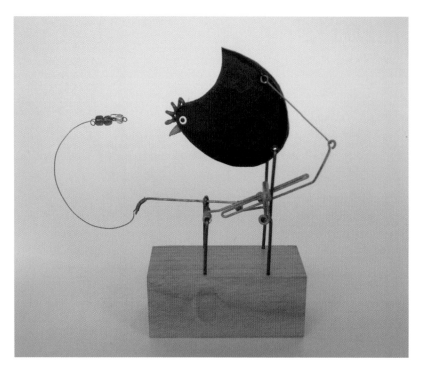

Ian McKay's *Big Hen*. The movements of the unconcealed mechanism are an important part of the overall effect.

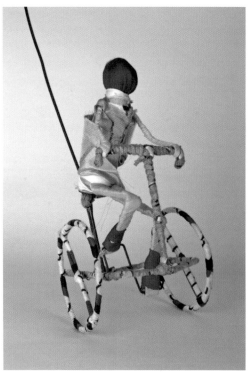

A Kenyan push-along toy. The rider seems to pedal furiously on his unlikely machine.

REVERSAL OF CAUSE AND EFFECT

Good use can be made of an illusory reversal of cause and effect. To give a simple example, the Kenyan toy cyclist is pushed along using a wire handle, and friction between the wheels and the ground ensures that the wheels turn. The axle is fixed to the wheels and turns also. It is a crankshaft with two cranks to which the legs are linked. They are hinged at hip and knee and, as the cranks turn, the legs move, flexing at the joints. In spite of the rudimentary modelling of the figure and the unlikely bicycle with two wheels abreast, the rather convincing effect is that the figure is pedalling away, and that it is the action of his legs that is turning the cranks, so turning the axle and the attached wheels, and moving the machine across the ground. It is possible to rotate a crankshaft through the reciprocal movement of an attached linkage (as is the case in the internal combustion engine, or in the wire helicopter illustrated in Chapter 3), but to actually achieve the movement of this toy through the action of the legs would require a much, much more complex mechanism.

In *A History of Aviation, The Age of Driftwood* two aeroplanes are spun around by pulling on a cord that runs round a pulley on the central shaft. The pulley is a loose fit on that shaft, the drive being transmitted through a ratchet and pawl so that when the cord is returned by a counterweight, ready for another pull (turning the pulley in the opposite direction) it does not impede the turning of the shaft. After a few pulls on the cord the planes move around fast enough to swing out on their wire supports, and, also, the propellers start to spin. On the real thing a motor turns the propeller, which pulls or pushes the aeroplane through the air, and flying models may use a twisted rubber band to the same effect. Here the reverse happens: the shaft rotates, carrying the aeroplanes round, and with them the propellers. As they move forward through the air their angled blades cause them to spin round.

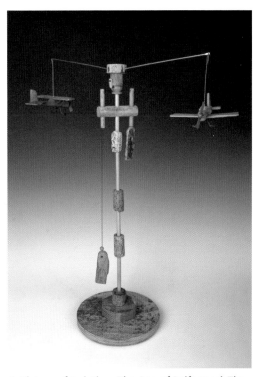

A History of Aviation, The Age of Driftwood. **The propellers spin as the planes *rotate*.**

BUY ONE, GET ONE FREE

The propellers are mounted so that they turn freely on a pin. This is an example of how effective it can be to attach further movable parts loosely to a piece that has been made to move with a mechanism. These extra bits will then also move, with no additional mechanism necessary. In the case of the aeroplane propellers the movement is quite regular, but in many cases the extra bits will move in an unpredictable way that makes the whole effect more lively. A comparison of two tin toys, a pair of pecking chickens and a pair of boxers, illustrates this clearly. Both consist of two figures mounted on a parallelogram linkage, and are operated by squeezing and releasing a springy hand grip. The chickens just move backwards and forwards as if pecking. The effect is regular, repetitive and not really very convincing.

Pecking Chickens. **A parallel linkage toy from China. The movement is rather stiff and mechanical.**

LEFT: *Boxers*. A parallel linkage toy from India. Loosely joined arms give a more dynamic and unpredictable movement.

BELOW: A wooden toy from Mexico, with a similar effect to the Indian tinplate boxers.

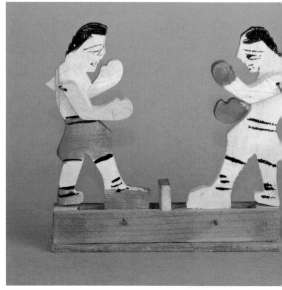

The boxers have essentially the same motion, but their arms are separate pieces, loosely attached at the shoulders so that they are free to rotate. As each boxer tips forward the front foot strikes the base, but the arms continue to move, swinging about in a variable and unpredictable manner. This is a much more dynamic and animated performance than that of the chickens.

Loosely attached arms are used to a similar effect on a pair of wooden boxers from Mexico. A different mechanism is used to move the bodies: the front feet of the figures are pivoted to the base, and linked by a flat steel spring. Pushing the spring down with a little wooden button brings the two figures towards the middle. When the button is released the figures spring up, their back legs strike the base, and the arms flail about.

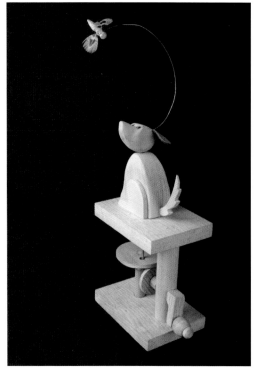

Dog and Insect. The unpredictable movements of the dog's head make the attached springy wire wobble, and the insect jiggles about.

In *Dog and Insect*, the dog's head is mounted on a metal rod that passes down within the hollow body and projects below through a hole in a wooden disc that rotates, through a friction drive, when the handle is turned. This causes the head to move about in an unpredictable way because of slight variations in friction both at the neck, where the head rests and pivots on the body, and also where the rod passes through the wooden disc. The insect is mounted on a fine springy wire attached to the head, so that as the head moves the insect moves with it. The reversal effect comes into play again here: it looks as if the dog is *following* the movements of the insect, rather than the reverse. But, in addition, the fineness and springiness of the wire allows the insect to bobble about – and to continue to do so for a bit when the handle is no longer turned.

Watching the Girls Go By

MAKE THEM LAUGH OR AT LEAST SMILE

In automata animate beings become machines and machines become animate beings. This more often seems absurd rather than scary or weird, but there is a residual edge of uncanniness. Possibly this explains why automata are such an appropriate medium for humour. They give you a touch of funny-peculiar along with the funny-ha-ha. The mechanical representation of human or animal movement is often funny in itself, and automata can present a visual joke in a sequence of events. A good title can help too: I make a driftwood piece which is a very simplified representation of a face: hat, eyes, nose, a rather rakish moustache, and nothing much more. A pendulum is hung from the nose. When it is swung the eyes look from side to side, and the moustache twitches. The title, *Watching the Girls Go By* contributes to its continuing popularity.

IDEAS, AND HOW TO GET THEM

Where do ideas for successful automata come from? They do not generally spring spontaneously into existence. A number of sources can be explored and strategies deployed to encourage them.

The Materials

One source of inspiration may be found in the forms and characteristics of the materials to be used. Take, for example *Thinks*, illustrated at the beginning of this chapter. This piece consists of the head and upper body of a pensive-looking character, who pushes down on, and then releases, a see-sawing track. A little widget rolls back and forth on the track, and he follows its movement with his eyes. The piece is largely made of driftwood, but the heart of it is the figure's torso, which is a substantial, but rusty and battered steel can. This had been partly flattened, pounded into crumpled folds by the sea, and deposited on the beach where I found it. The general shape without any alteration, immediately suggested a torso, with the folds being clothing, possibly monkish robes, and demanding the addition of a head and arms. I took it home, but had no immediate idea how to use it. Meanwhile I had finished a piece called *Man and Marble*, illustrated in Chapter 6. In that piece a see-sawing marble track cut from a large bamboo is supported on a steel shaft from the body, and directly moved by the mechanism. The arms move only because they are tied onto the ends of the track. This, incidentally, is another example of the reverse effect. It does look as if it is the movement of the arms that moves the track, and not the other way round. Looking for a variation on the theme, I came back to the rusty steel torso, giving it a head and arms, and mounting it so that the weight of one forearm and hand would push down a see-saw track at one end (it being counter-weighted at the other end). A shaft running across the front of the figure was fitted with a crank at one end to raise and lower the forearm, and with a cam at the other end operating through a wire linkage to move the eyes from side to side. To complement the combination of rusty steel and driftwood, the worm gear used to turn the shaft is made from a piece of old rope wrapped around a dowel.

Another example of taking inspiration from the materials is *Last of the Moccasins*. This piece was made for an exhibition in the cloisters of Salisbury Cathedral entitled *Made to Last*. A pair of shoe lasts was distributed to a number of different makers to work with. The foot shapes

Detail of *Thinks*, showing the crank to raise the arm, and the worm gear made from rope wrapped round a dowel.

Indonesian pull-along horserace toy.

suggested walking to me, but the sleek lines also suggest cars or carriages. The combination led in the direction of mounting a figure in each last, as if in a chariot, and finding a mechanism to move them up and down as they moved forward, as if walking. This requirement brought to mind a series of pull-along wheeled toys from Indonesia that I have in my collection, in which two cranked axles are linked by flat silhouettes of riders on horseback.

As this toy is pulled along, the two riders alternately rise forwards and fall back, as if continually exchanging the lead in a race. This mechanism adapted well to a combination of chariot race and walking feet. As the figures mounted on the lasts move along, and up and down, the arms move about a bit, independently, because they are rigidly attached to a shaft that passes through the shoulders and is free to rotate. The raised arm is counterbalanced by the lowered one. This is another 'buy one, get one free'.

Last of the Moccasins, my contribution to an exhibition of works made from shoe lasts.

Bird and Bee. The falling weight is halted twice per revolution by a wrapping escapement.

The Mechanisms

Another starting point for ideas may be the mechanism. This was the case with *Bird and Bee*. I first came across the wrapping escapement in Aubrey F Burstall's *Simple Working Models of*

No Fishing No Smoking uses a wrapping escapement and a small plastic clockwork motor. It was made for Robert Nathan, the manager of the British Toymakers Guild on his retirement.

Historic Machines. In a clock using this escapement a ball hanging by a thread from a rotating arm swings out and is guided by a 'wrapping arm' onto a 'wrapping post'. The falling weight driving the clock is stopped while the thread winds around the post, unwinds, winds round the other way, and unwinds again. The ball then swings round to a second post on the other side, and the process is repeated. In *Mechanical Toys, How Old Toys Work,* by Athelstan and Kathleen Spilhaus (see Chapter 6), there is an illustration showing a Japanese wind-up tin toy from around 1930, which uses this mechanism in a simplified form. There are four wrapping posts, but wrapping arms are not used, with the result that the thread only wraps around each post one way, before moving on. I subsequently found *Billy the Fisherman,* a British tin and plastic toy from the 1950s, illustrated in the introduction. This is a further simplification in having only one wrapping post. The rotating arm has become a fishing rod, and the ball on the thread a fish. I wanted to use this intriguing mechanism, and, taking inspiration from *Dog and Insect*, settled for a bird and bee, attaching a wire to the head of the bird for the rotating arm, and hanging the bee on a thin thread from that. It is driven through a thicker thread passing round a pulley on the bird's leg and over another pulley on each side. There is a weight permanently attached to the thread at one end, and at the other there is a hook on which a heavier weight is hung. The bird is pulled around, stopping twice in each revolution as the thread with the bee wraps round each post. When the heavier weight reaches the bottom of the drop, the hook tips up, and the weight falls off automatically. The permanent weight then pulls the bird in the opposite direction, stopping twice in each revolution as before, until the hook is at the top again.

The original version, with the 'wrapping arm' would not work in both directions like this, because the arm has to be fitted slightly to one side of the wrapping post. I also made a wind-up version of *Bird and Bee*, mounting the bird directly onto the shaft of a small clockwork motor from a plastic toy. I took more direct inspiration from *Billy the Fisherman* for a similar piece, *No Fishing No Smoking*.

The Imagery

In many cases the initial idea for an automaton comes from the possible subject matter, with the choice of materials and mechanisms feeding into and modifying the original concept as the design process proceeds. *Rat Race* is a complex piece that evolved from imagining a reversal of roles for laboratory rats and human experimenters. I pictured people running a maze, under the control of rats. I saw the rats looking down from above, as people do on rats, and this led to the idea of a prison, with the rats in observation towers. The maze became a circular track, suspended from the 'arms' of the rats in the towers,

Rat Race (detail).

with the 'human' figures scurrying around on wheels. This gave the possibility of moving the wheeled 'human' figures around by raising and lowering the rats' arms, and hence the track, in a sequence of movements controlled by cams. After some experimentation I settled on two towers and a central pillar for the rats, and two figure-of-eight tracks, one above the other. This was visually more exciting, but considerably complicated the construction, and introduced new problems, such as ensuring a smooth passage over the crossroads on the tracks.

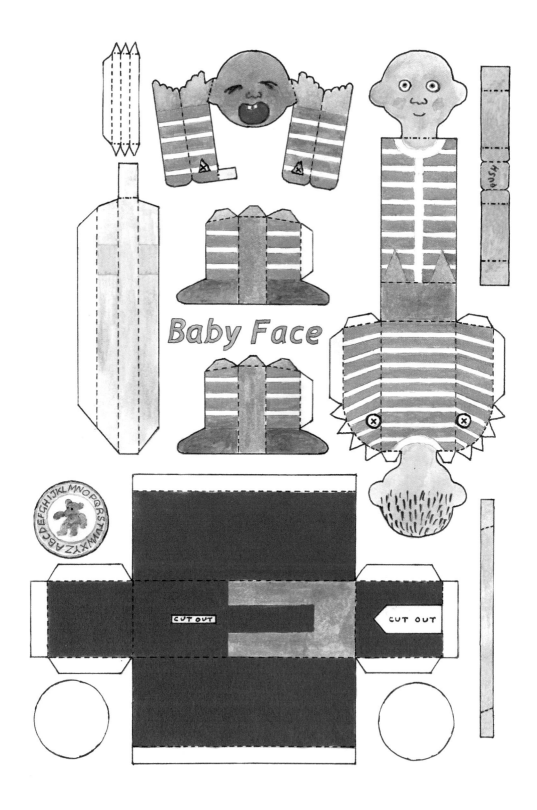

Baby Face

CUT OUT

CUT OUT

THREE PROJECTS

Baby Face
Signalling for Beginners
Swimmer and Shark

In this chapter, three projects will be followed through, from initial ideas to finished piece.

BABY FACE

The first project is a card cut-out with a simple mechanism. The inspiration for _Baby Face_ was _Cat with Mask_, illustrated in Chapter 1.

The essential effect of _Cat with Mask_ is the sudden change of feline expression from slightly gormless quietude to predatory menace. The arms holding the mask pivot on a wooden shaft through the shoulders. When the arms are raised the mask covers the face. This is achieved by pressing a button that pushes down a lever, housed in the base, which is attached to the back of one arm with a strong thread. When the button is released the arms and mask fall back down again under gravity. My initial idea was just to translate the original from wood into thin card, but I wondered if the same quick-change effect could be used for another subject. It struck me that the human baby might be a suitable subject, with his or her ability to change instantly from contentment to furious rage (and back again slightly more slowly). In addition, I wanted to produce a design that would fit on a single page.

Thin card is a versatile medium. On its own it can be cut, scored and folded, joined with slots and tabs, and built into complex 3D structures. With glued joints the range of possibilities is even wider.

One of the problems to be solved in translating a design from wood into thin card is how to make a rigid 3D structure from a rather flexible 2D sheet. Take for example a cube: it would be possible to cut six separate squares of card, and join them edge to edge to make up the cube. But this would not take full advantage of the properties of the thin card, which can be creased and folded without losing much strength. A better plan is to draw out the six square faces on a flat sheet, leaving as many edges as possible intact, but in an arrangement that can be folded into a cube. Such an arrangement is called a net. It turns out that there are eleven different possible nets for a cube (discounting rotations and reflections). They all have five edges to fold, and seven that are split in two, to give fourteen edges around the perimeter of the net. If a tab for gluing is added to every other one of these fourteen edges, then the net can be cut and scored, folded, and glued into a cube.

In designing a cut-out, one task is to divide the whole structure into a number of convenient elements, for each of which a 2D net can be drawn. As with the cube, a number of different nets may be possible. Two things to bear in mind in choosing which net to use are firstly that it

OPPOSITE PAGE: _Baby Face_ cut-out.

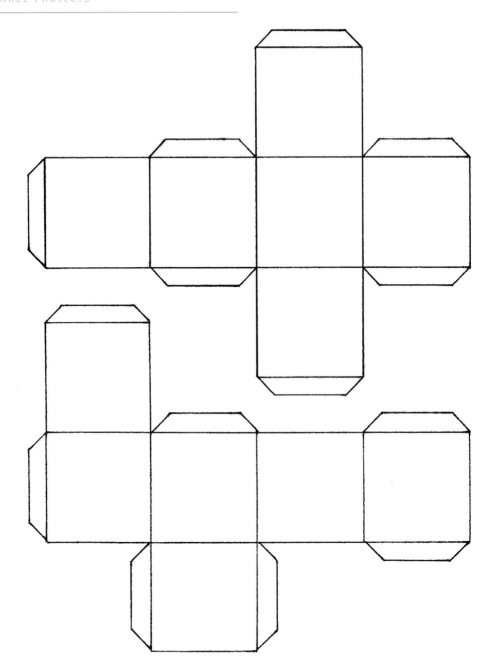

Two of the possible eleven different nets for a cube.

is good to achieve an economical layout of the different pieces needed to make up the whole cut-out, and secondly that folded edges will be slightly easier to keep neat than glued ones.

A number of changes have been made in converting the wooden *Cat with Mask* into the card

Baby Face. The cat exploits the colour of the wood – the body is yellow pine, which has a strong grain pattern. Paint is only used for the facial features. Plain white card is a bit dull, so the baby is painted.

The cat's body is pinned and glued to the *back* of the base – using wood this is simple to assemble and quite strong. The baby's body has been made a little thicker and glued to the *top* of the base. This makes it easy to add legs, which do not really feature in *Cat with Mask*. The baby's body is shaped so that it leans forward slightly. My first attempt had it upright, as in the cat, but I found that the arms, plus the plate/angry face that replaces the cat's mask, tended not to fall back easily when the button was released, even with the added weight of two extra layers of card between the plate and the angry face. Shaping the body so that it leans forward slightly solved this problem.

To keep things simple the arms and the head of the baby are made up of two layers of card stuck back to back, leaving them 2D.

In *Cat with Mask* the solid wooden base has a slot cut in it to house the lever, which is a strip of wood, rectangular in section, with a hole drilled through it near the front end, so that it can pivot on a thin round dowel. In *Baby Face* this lever is replaced with a piece of card scored, folded and glued into a slim, long piece with a triangular cross-section. This triangular cross-section gives the lever the

necessary strength and rigidity. A protruding tab is left at the front end to form a flexible hinge, to be glued into the hollow card base.

A thin round dowel also runs through a hole in the shoulders of the cat to join the two arms. In the baby this is replaced with card, again made up into a triangular cross-section. Little tabs at each end are used to glue this cross-shaft to the arms.

Making *Baby Face*

To make *Baby Face* you need access to a scanner, and a printer. Scan the page with the layout of the required pieces and print it on thin card (160gsm, 60lb cover stock, is good). The illustrations show it being made the same size as it is printed in the book. This works fine, but it is a bit fiddly, so you may prefer to enlarge it to fill an A4 sheet.

The more carefully you cut, score, fold and glue the better your cut-out will work. It is best not to hurry. Make sure you have read these

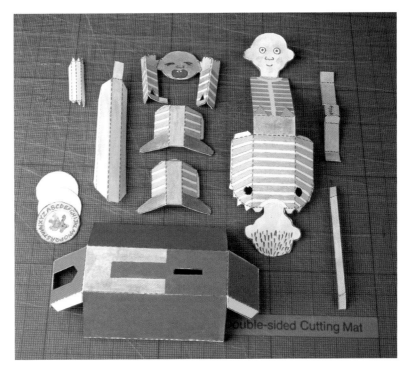

The pieces, cut and scored.

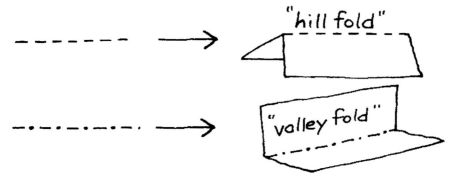

Hill folds and valley folds.

instructions and understand what to do before you start on each stage.

For cutting out the pieces of card you will need a sharp pair of scissors and a small craft knife. The kind with a retractable break-off blade is good, so that you can always work with a sharp point. For scoring the dotted lines, to make the folds, a *blunt* knife, such as an unserrated table knife, is good, or you could use the point of an old ballpoint pen

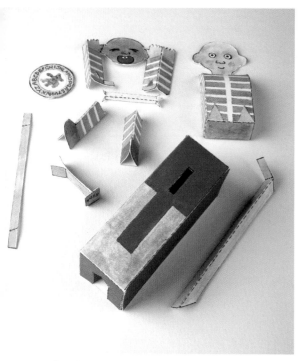

The pieces, partly glued.

that is completely out of ink. The aim of scoring is to press a narrow groove into the card without scratching or cutting through the surface. For scoring, and for cutting straight lines with the craft knife, use a cutting mat, or a piece of thick card, to rest the work on. Use a metal ruler as a guide.

Score all the dotted lines and cut out the internal holes in the pieces. There are two holes in the base (labelled *cut out*), two triangular holes in the arms, and two round holes in the shoulders (marked with a cross). I prefer to do this before cutting out the pieces.

Cut out all the pieces and fold on the dotted lines. The lines of dashes are hill folds, and the lines of alternate dots and dashes are valley folds.

For gluing, use white PVA wood glue, applied sparingly. Use a cocktail stick, or a scrap of card, to apply it. A clear glue, such as Bostik Clear or UHU, also works well.

Form the base into a box, open at the bottom. When the piece is fully assembled the square-ended tabs at the bottom of the base can be used to stick it onto a thick piece of card, to make a firm mounting.

Form the body by gluing the tabs at the bottom, then the long tabs at the front. Let the glue dry. Now apply glue to the small triangular tabs on the curved part of the body, and to the head. Carefully bring together and match up the two parts of the head, and the curve should fall into place over the triangular tabs.

Glue the two plain discs behind the teddy bear plate for extra weight.

For the yellow lever, the small white cross-shaft, and the two legs, apply glue to the long tab and form the triangular shape.

Glue the push button into a 'T' shape, leaving the two tabs at the bottom free.

Glue the tabs at the bottom of the 'T'-shaped push button in place (marked with darker yellow) on the long yellow lever.

Push one end of the white cross-shaft through the triangular hole in the left arm (the one without the yellow tab). Fold out the little triangular tabs, and glue the two sides of the arm together, sandwiching the little tabs so that the cross-shaft is attached to the arm, and sticks out at right angles from it.

The yellow lever now has to be glued in place inside the base. Put it inside the base through the hole cut at the back and then manoeuvre the push button through the narrow slot in the top of the base by temporarily folding the cross piece of the 'T' over to slip it through. Glue the hinge tab on the lever in place at the front of the base. It should be positioned centrally, with the end of the tab just reaching the bottom of the base.

Feed the cross-shaft, which you have already attached to one arm, through the holes at the shoulders of the baby. Check that the shaft turns freely. If it sticks try gently enlarging the shoulder holes with the point of a pencil. Feed the cross-shaft through again, and through the triangular hole in the right arm. As with the other arm, bend out the little triangular tabs and glue the two sides of the arm together, sandwiching the tabs. Now glue the teddy bear plate, with the two extra discs, onto the back of the angry face. Check that the arms can move up and down freely.

Now glue the baby's body in place on the base. When the glue is firm glue the two legs in place.

The last piece to attach is the narrow yellow strap, which joins the operating lever to the arms, replacing the thread used on the cat. Glue the tab at one end to the flat lower surface of

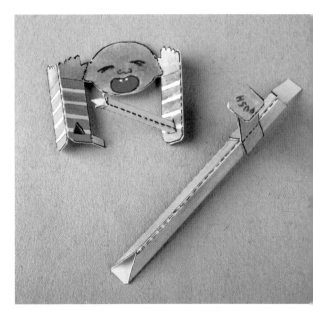

The lever, and the arms partly assembled.

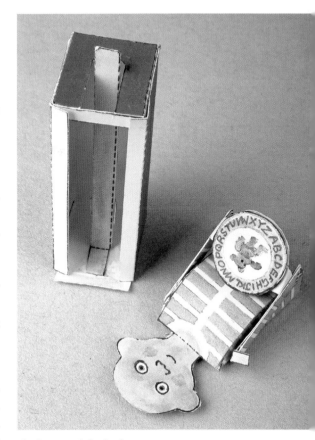

The base and the body.

Cat with Mask and **Baby Face** from the back.

the yellow lever, at the back. Line up the fold in the strap exactly with the end of the lever so that the strap is at an angle, pointing up towards the yellow tab at the back of the right arm. Finally glue the tab at the top end of the strap to the tab on the arm.

While gluing this tab it is important to push the lever down as far as it will go (using the push button), while at the same time holding the angry face up, tight against the smiley face. Hold everything in this position and glue the tabs so that the strap is just taut. When the glue is dry, test out the effect. If the angry face does not go up far enough when you push the button, or the teddy bear plate does not drop down far enough when you release it, then you may need to adjust the length of the strap. If necessary you could cut a fresh one from a scrap of card.

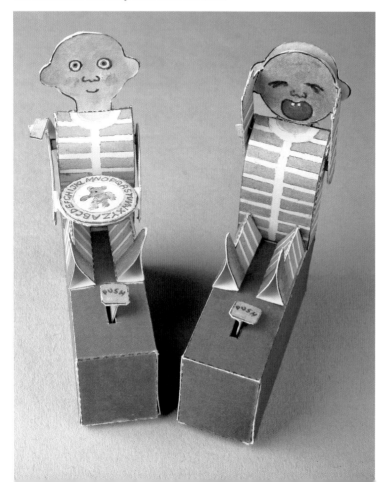

Baby Face completed.

SIGNALLING FOR BEGINNERS

The second project is an exercise that uses three differently shaped cams to control two figures that move their arms and turn. These figures are based on a traditional folk toy, the jumping jack, found for a very long time all over the world. Herodotus described an obscene example he saw in Egypt, and there was a brief, but intense, craze for them in eighteenth-century France.

Signalling for Beginners.

Typically in the jumping jack, or the French *pantin*, the limbs are attached to the body, but free to pivot, and strings attached to them at shoulders and hips are brought together into one string below. Pulling that string sets all the limbs dancing.

The figures in *Signalling for Beginners* have no legs, and the arms are waving flags. On the flags one figure has the letters HL, the other EP. At one point in the sequence, controlled by the cams, the figures turn, their arms cross, and the flags spell out the message 'HELP'.

For each figure a cam controls the arm movements, and a third makes the two figures turn together, because they are connected with a parallelogram linkage. If the camshaft were turned directly the sequence would happen rather too quickly, so a separate shaft, with a crank handle and a worm gear, is mounted at right angles. The worm gear turns a cogwheel on the camshaft. A similar drive, on a larger scale, is used in *Thinks*, illustrated in Chapter 4. There the cogwheel is cut from plywood, using a template made with a gear generator programme, and the worm gear is made by wrapping rope of a suitable thickness around a dowel. On this smaller scale ready-made gears are used. Brass gears, such as the vintage Meccano used in the model of the Chinese *South pointing chariot*, illustrated in the introduction, would be good, but relatively cheap plastic mouldings have been used here. See Chapter 6 for suppliers.

Making Signalling for Beginners

The figures

The body of each figure is simply a rectangular piece of wood, 65 x 35 x 12mm (2^1/2 x 1^1/4 x ^{1}/2in), just rounded off slightly at the shoulders. An oval head is cut from the same material. Holes are drilled to take a two 6mm (^{1}/4in) dowels: one for the neck about 20 mm (^{3}/4in) long, and one to support the body about 80mm (3in) long. The arms are 12mm x 70mm, and about 5mm thick (^{1}/2 x 2^3/4 x ^{3}/16in). At the hand end a 2mm (^{5}/64in) hole is drilled at about forty five degrees, to take a piece of cocktail stick with a paper flag glued to it. The left-hand figure has H and L, and the right-hand figure has E and P, so that when they face forwards and raise their arms the flags spell out HELP. Looked at from the back the flags spell out PLEH in the international code of signals. Well, they are beginners.

At the shoulder end, drill two holes. The position of these holes is important: the first hole is made centrally about 10mm (^{3}/8in) from the end. It should be a loose fit for a small nail or gimp pin, used to attach the arm to the shoulder. This nail should have a distinct head so that the arms are free to rotate without falling off. A second hole, to take the operating thread is made about 3mm (^{1}/8in) from the shoulder end, and towards the *outer* corner when the arm is down. If the holes for the thread are not so placed the arms may not go far enough up when the thread is pulled down.

Two extra pieces 50mm (2in) long are cut, from the same material as the arms, for the levers, which are mounted on the dowels below the bodies. These form part of the parallel linkage that joins the two figures to make them turn together. Each has a hole drilled centrally that is a snug fit on the dowel, and a small hole about 5mm (^{3}/16in) from each end to take the wires that will complete the parallel linkage. Another shorter piece, 25mm (1in) long is also needed, which will be attached to the dowel support of the left-hand figure, below the top of the framework, to attach the wire that links it to the cam follower. It has a hole near one end to fit snugly on the dowel, and a small hole near the other to take the wire.

At this stage the arms can be attached at the shoulders with the small nails, far enough apart to allow the arms to rotate without the ends overlapping in the middle. A strong thread is fed through the other holes and tied with a thumb knot, making a loop that joins the two arms. One end is trimmed short and the other needs to be plenty long enough to pass through a hole in the top of the framework and be tied off to a screw eye on the cam follower. Pulling on this string should raise the arms up above the heads of the figures. Glue the dowels into the bodies and heads, and push the levers for the parallel linkage into place on the dowels below the bodies, but do not glue them yet, because their angle to the bodies will need adjusting later. The remaining short lever will be fitted to the dowel below the figure on the left when it has been mounted into the top of the framework.

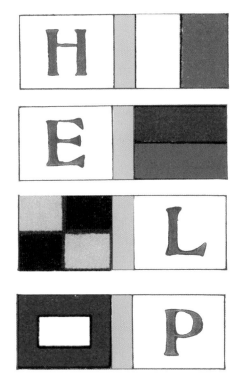

The flags.

The figures.

The framework

I used some oak offcuts, joined with wood screws, so that dismantling for fiddling about with the mechanism is easy. You can vary the material, method of fixing and dimensions of the framework, but if you follow the dimensions given here for the other parts, the figures should be mounted in the top, 90mm (3^1/$_2$in) apart. The holes for their plastic guides, and for the strings from the arms to pass through, should be on a diagonal line from the front right to the back left corner of the top. The gap between the top and the base should not be less than 80mm (3in) to allow room for the cams and followers. Here the top and the base are 180 x 65mm (7 x 2^1/$_2$in). The top is 8mm (5/16in) and the base is 30mm (1^1/$_4$in) thick. The side supports, 110 x 30 x 12mm (4^1/$_4$ x 1^1/$_4$ x ^{1}/$_2$in), are set into the ends of the base, each secured with a single screw. The top is screwed down onto them. To support the figures two plastic tubes, 25mm (1in) long, cut from a dead ballpoint pen casing, are mounted in the top. Choose a pen casing that is a good match with the dowel, allowing it to rotate freely, but holding it upright. The holes for mounting these guides need to be matched to the outer diameter of the pen used. Immediately to the left of each of the holes made to take these plastic tubes is another hole, about 8mm (5/16in) diameter, for the string from the arms to pass through.

The two supports for the cam followers, 50 x 20 x 8mm (2 x 3/4 x 5/16in), and the two supports for the drive shaft, about 65 x 20 x 8mm (2^1/$_2$ x 3/4 x 5/16in), are simply screwed to the outside of the solid base block. One support, fixed to the front of the base, takes two cam followers, the one that raises the arms of the left hand figure on the left, and the one for turning the figures on the right. The other support, fixed to the back, takes just the cam follower that raises the arms of the right hand figure. These supports must be positioned on the base so that the wooden cam

followers controlling the arms lie immediately below the holes in the top of the framework through which the arm strings come down.

The camshaft

The camshaft carries the three cams, and the plastic cogwheel through which it is driven. A 6mm (^{1}/$_4$in) dowel is used, that runs through holes in the end pieces of the framework. These holes (40mm (1^1/$_2$in) from the top) should be large enough for the shaft to turn easily, but if they are too loose, you will have problems meshing the cogwheel with the worm gear on the drive shaft. Leave enough room at each end of the shaft to glue on a wooden bead outside the end pieces of the framework, stopping the shaft from moving from side to side. You can make a simpler version by leaving out the plastic cog wheel and the separate drive shaft altogether, and replacing the bead at the right with a wooden crank handle to drive the camshaft directly. As mentioned before, this will make the sequence happen rather quickly, unless the handle is turned very slowly.

The cams and followers

The cam followers are wooden levers that are mounted on supports screwed to the base of the framework. They are cut from wood about 8 x 5mm (5/16 x 3/16in) in section. The two for the arms are 65mm (2^1/$_2$in) long. Each has a hole at one end that is a loose fit on a small screw used to attach it to its support, and at the other end a small screw eye is fixed for tying the thread. In between, at about 40mm (1^1/$_2$in) from the screw end, drill a hole to take a projecting metal pin that will slide against the edge of its cam as it rotates (the wooden part of the follower lies just to one side of the cam). For the two cams controlling the movement of the arms, the follow-

RIGHT: **The three cams. The central one turns the figures, the other two raise the arms.**

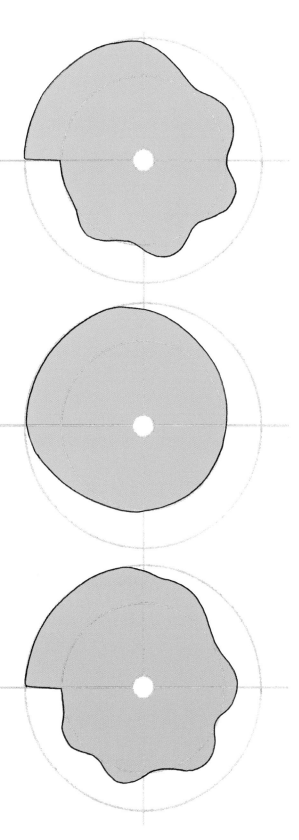

ers are *below* the cams. When attached with the thread, the weight of the arms and flags pull the follower up against the cam, so that the metal pin follows its contours. For the cam controlling the turning of the figures the follower is to the front. This is similar to the other two, but a little longer, 75mm (3in). It has a hole for the screw to attach it to the support at the bottom, and at the top a small hole to take a wire link to the short lever on the dowel supporting the left-hand figure. The metal pin is at about 40mm (1 1/2in) from the screw end. This follower will need a spring to hold it back against its cam. You need a really light spring, and it is sometimes difficult to find, or make, a metal spring that is appropriate. One simple solution is to use a rubber band instead. The disadvantage is that rubber degrades after a while, losing its elasticity, especially in strong light. On the other hand, rubber bands come in a wide range of lengths and strengths, and are easily and cheaply replaced. Attach the rubber band to screw eyes, one in the base, and one in the follower, placed between the screw attachment and the projecting metal pin that bears on the cam.

The shape of the cams controls the movement of the figures. The sequence is that the two figures begin looking to the right, waving their arms at random. They turn ninety degrees, towards the front, and raise their arms together. The arms in the middle cross, and the flags spell out 'HELP'. They then drop their arms together, more suddenly, and turn back towards the right. The sequence then repeats. The strings from the arms need to be pulled down about 15mm (5/8in) to raise them, and a similar pull on the short crank at the bottom of the shaft of the left hand figure will turn the figures through the ninety degrees required. Each cam follower is a Class three lever (*see* Chapter 3) with the fulcrum at one end, the load at the other, and the push

The base of the framework, with the drive shaft, the cams and the followers fitted.

from the cam on the metal pin somewhere in between. With the measurements given above, the end of the followers will move about one and a half times as far as the cam pushes the metal pin. This means that 10mm (³/8in) will be enough for the throw of the cam. That will give the required movement of about 15mm (⁵/8in) at the end of the follower.

To make the shapes for the cams, a circle, 65mm (2¹/2in) in diameter, is drawn, which should fit comfortably in the framework, leaving room for the followers below. To give a throw of 10mm (³/8in), a second concentric circle 45mm (1³/4in) in diameter is added. The profile of the cam lies between these limits. The cam to turn the figures has a quarter at the minimum diam-

eter (the figures will face right, looking from the front), and a quarter at the maximum (the figures will face forwards). In the quarters between, a smooth curve makes the transition (the figures will be turning). The cams to lift the arms have a quarter at the maximum diameter ending with a radial line straight back to the minimum (giving a period with the arms raised, followed by an abrupt drop). Cams with this pattern are known as 'snail cams'. Such cams only work in one direction. For the other two quarters a wobbly curve is drawn, finally reaching back to the maximum diameter. The bumps are drawn slightly differently in the two cams (giving a period when the figures wave their arms up and down a bit, not in synch).

The cams have been cut in 9mm ($^3/_8$in) ply-wood using an electric fretsaw (scroll saw). A hand fretsaw, a coping saw or a piercing saw will also do the job (with a little more hard work). If the hole drilled to mount the cams on the dowel shaft is a good tight fit you can push them onto it, adjust their position, and test the movement. Be careful to get the three cams the right way round, and in the right order. At the final stage fix them by marking their position, sliding them slightly to one side, applying a touch of glue and sliding them back. If the fit on the shaft is a bit loose, or if the cams are cut in thinner material, it is best to cut a wooden collar, or hub, to attach to each cam, which will provide a larger surface for the glue, and if necessary it can be fixed to the dowel shaft with a small screw or panel pin.

The plastic cogwheel used here also needed the central hole drilling out to fit the dowel shaft.

The drive shaft

The plastic worm gear used here is too slim to fit on a wooden shaft. You need a fairly rigid metal wire or rod. Here the central hole in the worm gear has been drilled out with a 2.4mm ($^3/_{32}$in) bit to fit a length of bronze brazing rod. This is about as thick as the worm gear could take. You could also use a slightly thinner, but reasonably stiff, wire if you make the hole in the gear accordingly smaller. The drive shaft is mounted, free to turn, in holes in the wooden supports, which are approximately 65 x 20 x 8mm (2$^1/_2$ x $^3/_4$ x $^5/_{16}$in). Adjust these carefully to the exact length required for the worm gear to mesh cleanly with the cogwheel on the cam shaft, before screwing to each side of the base. Feed a short piece of plastic tube onto the shaft on each side of the worm gear, to hold it centrally, and when you have screwed the supports in place, trim the rod off close to the support at the back, and bend to form a crank handle at the front.

Assembly and adjustment

The two figures are mounted in the top of the framework, in the guides made from a plastic pen casing. The wooden levers mounted on the dowels below the figures rest on top of these plastic guides.

When the top of the frame has been screwed in place these levers must be linked together using two thin wires 110mm (4$^1/_4$in) long, bent at right angles 10 mm ($^3/_8$in) from each end, making two 90mm (3$^1/_2$in) links. Brass wire 0.8mm ($^1/_{32}$in), which can be found in model shops (see Chapter 6) has been used here, but galvanized steel wire, 0.9 mm diameter, or similar, would be fine. Now glue the levers in place on the dowels supporting the figures. They are in the correct position when the parallel linkage forms a rectangular shape with the figures at the half way point of their ninety degree turn, facing forty five degrees to the right (from the front). This ensures that the figures will be able to turn fully to the side, and fully to the front.

Now tie the string from the arms of each figure to the screw eye on the cam follower. Feed the string through the screw eye, turn the camshaft so that the widest part of the cam bears onto the metal pin on the follower, pull the string so that the arms are at their high point and tie off. Fit the rubber band spring to the middle cam follower and link that follower to the little crank arm at the bottom of the dowel from the left hand figure with a piece of the thin wire about 65mm (2$^1/_2$in) long, with bends 15mm ($^5/_8$in) from each end, at right angles to each other. Push the ends through the hole in the little crank arm, and in the cam follower, and bend again slightly to stop the wire becoming detached. The little crank arm may need rotating on the dowel until you find the position where, when the camshaft is turned, the figures face to the right (from the front) at one extreme, and forwards at the other. Then glue it in place. Finally, rotate the cams on the shaft to line them up correctly. Align the

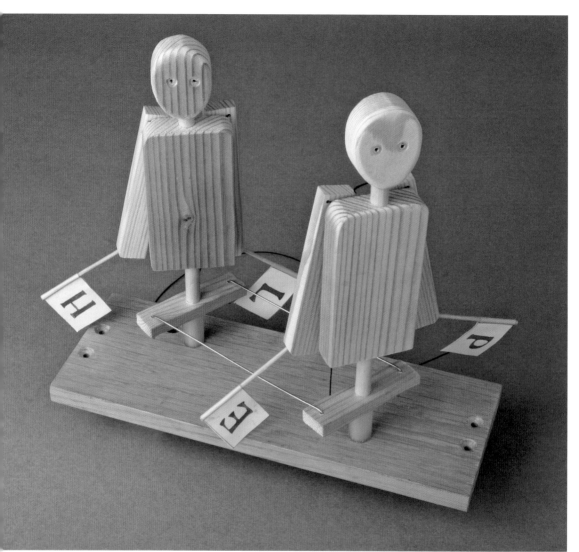

The figures mounted on the framework top, with the brass wires completing the parallel linkage.

two cams controlling the arms so that the arms drop suddenly together. At that point the figures should be on the point of turning back towards the right. For this to happen the middle cam follower must be just at the end of the wide part of the cam. When the movement is right, with the figures turning to the front and raising their arms to spell out the message 'HELP', mark the position of the cams, and glue them in place.

SWIMMER AND SHARK

The third project has an African inspiration: I have a number of push-along toys from Zimbabwe that depict a human figure chased by a dangerous animal.

The figures are in painted wood, and tinplate, and the toys all have a similar mechanism made of steel wire. The wooden wheels are fixed to a wire axle that is bent to form cranks. As the toy is pushed along the ground, using a long wire handle, the wheels and the axle turn, and wire linkages convey movement from the cranks to the various moving parts of the figures.

This design takes the imagery of one of these toys, a swimmer pursued by a shark, but replaces the wheels with a crank handle. For the swimmer it closely follows the original in the ingenious use of bent wire in attaching, articulating and moving the limbs. For the shark I have introduced a variation. In the original, the body is rigid, with just the tail moving from side to side. This works pretty well since the push-along toy actually moves forward, making the pursuit dynamic and convincing.

Exit, Pursued by a Hippo. **A Zimbabwean wood and wire push-along.**

In the static version I felt that giving the shark a slightly more sinuous swimming motion would compensate for the lack of actual forward movement. There is therefore an additional joint in the body of the shark, allowing it to flex.

Push-along swimmer and shark.

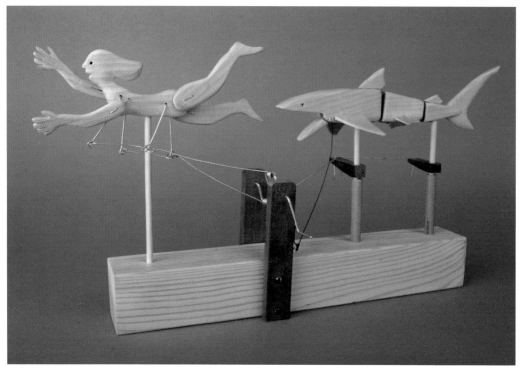

Swimmer and Shark.

Making *Swimmer and Shark*

The base and supports

The starting point is a solid base. For small automata operated with a hand-turned crank it can be slightly annoying if you have to use your other hand to hold down a lightweight base or framework. Any sufficiently chunky piece of wood will do here. I used a piece of pine from a discarded bedframe, 270 x 40 x 40mm (10½ x 1½ x 1½in). Having opted for a solid base you need to add a support for the wire crankshaft. This consists of a pair of wooden uprights screwed to either side of the base block. I used two scraps of hardwood 90 x 16 x 8mm (3½ x ⅝ x ⁵⁄₁₆in). The next step is to provide the supports for the swimmer and shark. The swimmer,

and the head and tail sections of the shark are each mounted on bamboo skewers, about 4 to 5mm thick (⁵⁄₃₂ to ³⁄₁₆in), cut to a length of about 12cm (4¾in). A wooden dowel of a similar thickness would be just as good. The skewer or dowel supporting the swimmer's body will be fixed firmly to the base, but the shark's head and tail have to be free to rotate a little. To allow for this, fix two tubes into the base to support the skewers, which will themselves be firmly fixed to the shark. These tubes must be wide enough for the skewers to fit into them easily, and turn freely. To make the tubes I used sections of bamboo about 60mm (2¼in) long, cut from a garden cane. Garden canes are relatively cheap and widely available. As mentioned in Chapter 1, they are hollow, except at the nodes, and taper along their length, so you can cut tubes of different diameters from them. They vary a good deal in the thickness of the outer wall, and the resulting inner diameter of the tube, so it is worth having a few to choose from. Drill three holes along the centre line of the base block. The first, 25mm (1in) from the end, is to take the support for the bather, and it must be a snug fit for the skewer. The second, 180mm (7in) from the same end, must be a snug fit for the bamboo tube that supports the head of the shark. The outer dimension of the bamboo that I used is approximately 8mm (5/16in). It is best to leave drilling the *third* hole, to take the bamboo tube for the tail, until you have assembled the shark, when you can measure the precise distance required between the head and tail supports.

In the push-along toy, the shark's tail, and the wire extension that sticks out from it at right angles, together form a *bell crank* (*see* Chapter 3), which converts the backwards and forwards motion of the link from the crankshaft into a side to side oscillation of the tail. The new version also uses bell cranks to move both the head and the tail of the shark, but the projecting arm of each crank has been moved down onto the vertical supporting shaft. I used small pieces of hardwood, about 50 x 8 x 8mm (2 x 5/8 x 5/8in) for these projecting crank arms. Each has a hole drilled through it near one end that is a snug fit on the shaft, and the other end is tapered a bit. The arm below the tail has a small hole near the end to take a wire link to the arm below the body, which has *two* small holes, one near the end for the wire link to the tail, and one about 10mm (3/8in) from the first for the link to the crankshaft. Don't glue the arms on at this stage so that you can adjust their position during the final assembly.

The wire crankshaft and links

The wire is galvanized steel. For the crankshaft I used wire 1.4mm in diameter (AWG15). This is easily enough bent and cut, with a pair of long nose pliers, and is just about thick enough at this scale. Any thinner and it would be liable to

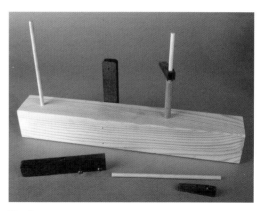

The base.

bend out of shape too easily. Wire a little thicker would be fine, but, of course, the thicker it is the more difficult it is to manipulate, bend and cut. For the rest of the wire work I have used wire 0.9mm in diameter (AWG19), which again, is just thick enough at this scale, and easier than the thicker gauge to form into the coiled ends of the links. It is not always easy to source the thickness of galvanized steel wire that you want, but something a little thicker would still be fine. One link on each crank will move the swimmer's limbs. The crank at the back has a second link, which will move the shark.

Make up the crankshaft, complete with links: the first task is to make the three links that must be added to the crankshaft as it is bent into shape. If you attach a wire link to a wire crank with a simple loop-eye at the end it will tend to slip around the elbow of the crank, and jam. One possible solution, used on the tinplate see-saw in Chapter 3, is to feed a sleeve onto the shaft on each side of the loop. Another is to wrap a thinner wire securely around the crank on each side of the loop, as in the wire helicopter also in Chapter 3. A more elegant alternative is to form a cylindrical coil at the end of the link, in place of the single loop. When the crank has been bent into shape around it the coil is securely positioned and cannot slip around the bend. The easiest way to form the coil is on a piece of wire slightly stouter than the crankshaft. Hold

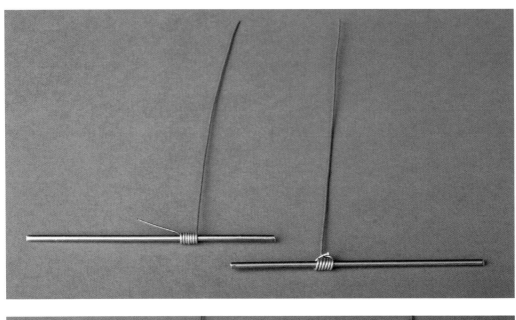

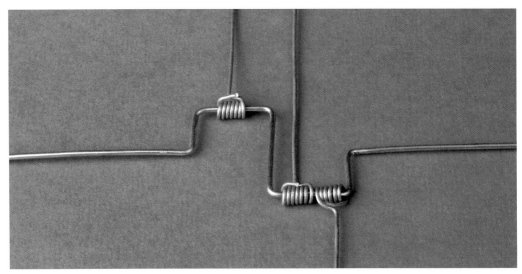

Forming the crankshaft and connecting rods.

the end of a piece of the thinner wire, about 20cm (8in) long, against this thicker wire and wrap it around half a dozen times or so, trying to keep each loop tight against the last. Bend the long end of the wire back towards the middle of the coil, and then out at right angles, while the short end is similarly bent towards the middle, and trimmed off with wire cutters (if you don't do this, the projecting end of the wire may catch on the crank).

You can then slide the completed coil off the wire former, and it should fit loosely on the wire used to form the crankshaft.

The three links formed in this way can then be fed onto the crankshaft as it is bent to shape. Starting with a straight piece of the 1.4mm (AWG15) wire, about 25cm (10in) long, a bend is made about 8cm (3in) from one end by gripping the wire with a pair of long nose pliers and pushing it firmly with the other hand, close to the pliers. The aim is to make a reasonably sharp right angle. Make the next bend in the opposite direction, forming one arm of the crank. The length of this arm determines the *throw* of the crank, the distance that the linkage attached to it will move back and forth as the shaft rotates. That distance will be *twice* the length of the crank arm. Here the arm is made about 12mm (1/2in) long, so the throw will be about 24mm (1in). Before you bend the wire back again to form the other arm of the crank, slide the coiled end of the first link onto the crank.

The next bend determines the *width* of the crank. There are two cranks on the shaft, and 40mm (1 1/2in) between the wooden supports, so each crank needs to be a little less than 20mm (3/4in) wide, to allow for free movement of the shaft, say 17 or 18mm (11/16in). To make this bend, grip the wire beside the coiled end of the linkage, using the pointy end of the long nose pliers. The bend should be in the same plane as the last one so that the arms of the crank are in line. If they are not, twist the wire a bit to line them up.

The crankshaft in place.

Make the next bend 24mm (1in) from the last, because the second arm of the first crank is extended to form the first arm of the second crank. The second crank is formed in a similar way to the first, except that *two* linkages are slipped onto the crank before the second arm is formed. The last bend brings the crankshaft back to the axis. For the shaft to rotate freely when mounted in its supports, it is important that the two end sections line up and form a straight axis – you may need to tweak a little to achieve this.

Now you can mount the crankshaft on the base. Unscrew one of the supports from the base, so that you can feed the ends of the crankshaft through the holes in the supports (the crank with two linkages to the rear). Screw the support that was removed back onto the base. At the front, bend the shaft at right angles close to the support and then back again, and cut it off at a suitable length to form a crank handle. Check that the shaft turns freely. At the back, cut off the shaft, leaving a small projection, which can be bent over. Cut the three links to the necessary length when the shark and the swimmer have been mounted on the base.

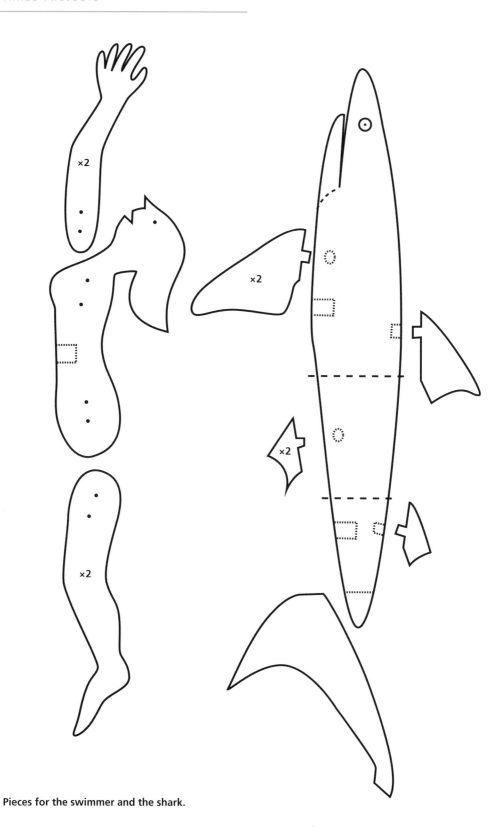

Pieces for the swimmer and the shark.

The figures

Lime wood, which is close grained and cuts and carves very well, has been used for the figures. However, since the pieces are basically profiles, and need only slightly rounding off, rather than intricate carving, any wood of a suitable thickness would do. The shapes have been cut using an electric fretsaw (scroll saw). A hand fretsaw, a coping saw or a piercing saw will also do the job (with a little more hard work). Further shaping can be done by whittling with a carver's knife and/or by sanding. The wood used for the bodies of the swimmer and the shark is about 10mm (³/₈in) thick. The arms and legs of the swimmer are about 5mm (³/₁₆) thick. These can be cut from a thinner board, but I actually cut the shapes of one arm and one leg from the 10mm (³/₈in) board, and these were then sawn down the middle to make pairs. Similarly the tail and fins of the shark can be cut from thinner material, or cut down from the 10mm (³/₈in) board.

The body of the shark can be cut as one piece, complete with the lower jaw. Cut the lower jaw off, and cut the body into head, middle and tail sections. These three sections will be re-joined

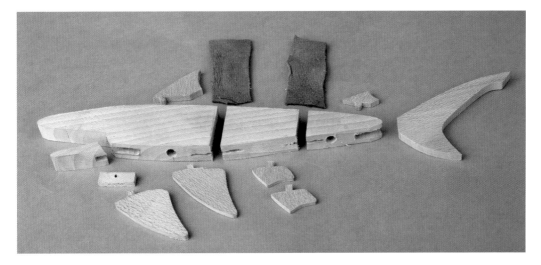

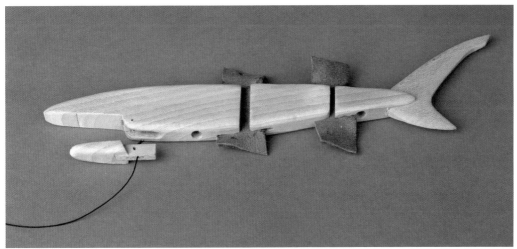

Stages in assembling the shark.

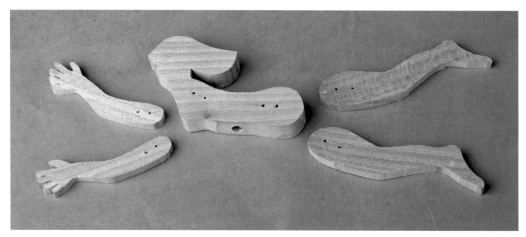

The swimmer cut out and drilled.

with hinges made from leather, glued into vertical saw-cuts. A piece of strong cloth – denim for instance – works well as an alternative. Ideally the leather or cloth should be thick enough to just slide easily into the saw-cuts. For each of the two joints in the body, trim the leather wide enough to fill both saw-cuts, plus an allowance of 4 or 5mm (³/₁₆in) between the pieces to allow the hinge to flex easily. Leave enough projecting above and below the body to hold onto while working the leather into the saw-cut. Try it for fit, and then remove the leather, which will be glued in place after shaping and sanding the body. You need to cut a slot in the tail section at the back, to take the tail itself.

To re-attach the lower jaw, cut a slot in the back of it, and then let a small tenon, about a third of the thickness of the jaw, into it and glue it in place. To house this tenon, make a slot in the head piece from below, behind the mouth. It can be made with a series of drilled holes, cleaned up with a small chisel or a knife. Drill a hole to match the thickness of the skewers in the lower surface of the swimmer's body and of the shark's head and tail sections.

Each limb of the swimmer also needs two small holes drilled from the side, the first about 6mm (¼in) from the shoulder/hip and the second 6mm (¼in) from the first. Ideally these holes should be quite a snug fit on the 0.9mm (AWG19) wire. Drill four holes through the body: two at the shoulder and two at the hip, as shown in the picture. These holes need to be a loose enough fit on the wire for it to rotate freely.

Now you can round off the pieces, by whittling with a carver's knife, and and smoothe them by sanding.

Assembling the shark

Apply glue in the saw-cuts and work the leather/cloth hinges into them. When the glue is dry, trim off the leather or cloth top and bottom.

Fit the lower jaw in place, with the little tenon in the slot, and drill a small hole right through, just behind the mouth, to take a piece of wire, which will form the hinge. Thread a wire through, and check that the mouth opens freely. Take apart, and drill another small hole vertically through the tenon, *just* behind the hinge point. This is to attach a thread that will eventually be fixed to the base to open and close the shark's mouth.

Drill holes, in the top, along the central line, to fit the little lugs on the dorsal fins. Similarly, drill holes on each side near the bottom, at a suitable angle to take the lugs on the pectoral and pelvic fins. Glue in all the fins.

The shark complete with lower jaw and fins.

Finally, glue the skewers into the head and tail sections, each with the projecting wooden crank arm roughly in place, but not yet glued. At this stage you can measure the exact distance between the head and tail supports, and drill a hole in the base to take the second bamboo tube (for the tail shaft) at that distance from the hole for the head support.

Assembling the swimmer

Each limb must be articulated to the body of the swimmer. Use a piece of the thinner wire that will also form the arm of a bell crank, projecting below the body. The arm and leg on each side will be linked together, and then attached to one of the links on the crankshaft. You need four pieces of the thinner wire about 10cm (4in) long: one for each limb. Make a right-angled bend about 10mm (³/₈in) from one end, and then another bend, in the same direction, at a distance that matches the distance between the two holes in the wooden limb. From the outside of the limb, push the long end of the wire right through the hole nearest the shoulder/hip until

the bent end of the wire reaches the other hole. Work it through that hole, pushing and pulling the wire through as far as it will go. On the inside of the limb, bend the short end back along the limb, towards the hand or foot, and bend the long end back so that it projects straight out beyond the shoulder or hip. Tap these bends down flat against the limb with a light hammer. Repeat for each limb. It is important to remember to make a right and a left pair of arms, and of legs, not two left or two right.

Articulating the limbs: taking the left leg, make a right-angled bend in the wire projecting at the hip, *level* with the end of the leg. Feed this wire through the *rearmost* hole in the body, and then, holding the leg in position against the body, bend down at right angles. This secures the leg in place, and forms a crank arm projecting below the body. Now add the right leg, using the remaining hole at the hip, but, because this hole is a little distance in front of the other, make the right-angled bend in the projecting wire *a little beyond* the end of the leg, so that the two legs end up level on the body. The projecting wire is bent down at right angles to secure the leg as before.

Attach the arms in the same way, attaching the right arm through the *foremost* hole in the body. The four projecting cranks are trimmed to about 40mm (1½in) and, using long-nose pliers, make a little loop-eye in the end of each. Finally, at each side of the body the arm and leg cranks are linked with a short length of wire with a loop-eye at each end. The length between the loop-eyes should be equal to the distance between the holes through which the arms and legs are mounted on the body.

Mounting and linking the figures

Mount the swimmer on its supporting shaft, and connect the arm and leg on each side to a link on the main crankshaft. Pull the wire projection from the right leg back, which raises the leg (and lowers the right arm), and rotate the main crankshaft so that the crank at the front is as far back as it will go. Then trim the linkage attached to the crank to about 6mm (¼in) longer than reaches to the loop-eye on the projection from the right leg, and another loop-eye is formed to

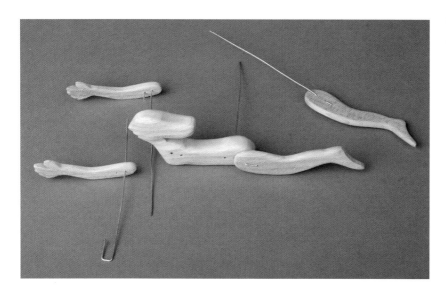

The swimmer. Articulating the limbs, stage one.

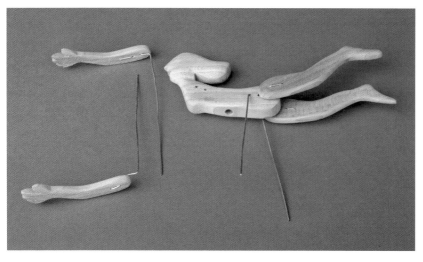

The swimmer. Articulating the limbs, stage two.

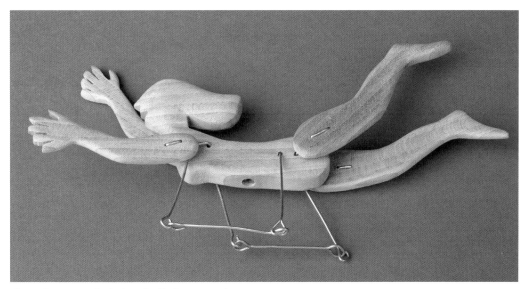

The swimmer. Articulating the limbs, stage three.

attach the two links together. When these loops have been linked together, turn the handle on the main crankshaft to check that the leg and arm move appropriately. You can bend the wire articulation a bit one way or the other at the shoulder or hip, so that the arm and leg are in a suitable position. Repeat the process to attach the left leg to the innermost of the two linkages on the other crank.

The shark: place the shafts supporting the head and tail sections into the bamboo tubes projecting from the base, and position and glue the wooden crank arms just clear of the bamboo tubes, projecting towards the back, at right angles to the shark's body. Cut a piece of wire 30 mm (1¼in) longer than the distance between the supporting shafts and bend at right angles, 15mm (⅝in) from each end. Insert these ends from above into the holes near the ends of the wooden crank arms, linking the head and tail sections, so that when the head turns the tail turns also. Now bend the remaining linkage on the main crankshaft at right angles and feed it through the second hole in the wooden crank arm on the head support. The position for this

bend is measured by rotating the main crankshaft until the crank is as far forward as it will go, and pushing the wooden crank arm forward as well. Test for a good swimming movement, and adjust the length of the wire link as necessary.

Finally, attach the string from the jaw to the base so that the side to side movement of the head section also opens and closes the mouth. Fix a small screw eye into the front support for the crankshaft, near the top of the base, and tie the thread to this, so that it is pulling the jaw closed when the head section is pointing back as far as it goes. This requires quite a delicate adjustment to get the optimum opening and closing of the jaws.

Final adjustment

On turning the crank handle everything should now move freely. The angle of the swimmer's arm and legs can be adjusted again for best effect by bending them a little up or down at the shoulders and hips.

USEFUL STUFF AND WHERE TO FIND IT

SOURCES

Educational suppliers catering for Design and Technology in schools are a good source of materials and of mechanical components, such as gears, pulleys and small motors. The polythene worm gear and cog used in *Signalling for Beginners*, the second project in Chapter 5, was from: www.mindsetsonline.co.uk, a not-for-profit company owned by Middlesex University. They supply all sorts of useful stuff at a reasonable price.

For making wooden gears on a slightly larger scale I use a gear generator program from www.woodgears.ca.

Useful wooden components, such as dowels, wheels and axle pegs, are available from www.woodcraftsupplies.co.uk, or www.craftparts.com in the US. If you are fortunate enough to have a local model shop, it will have small tools and adhesives and will probably stock materials such as wooden dowels, and metal wires and tubes, for example, those from K & S Metals (www.ksmetals.com). Otherwise searching the internet will usually reveal a source.

Kebab skewers, both birchwood and bamboo, are widely available in supermarkets, but Chinese supermarkets and kitchenware suppliers usually have a broader range.

Specialist outlets such as fishing tackle shops are a good source of lines and threads, swivels, guides, weights and other unusual bits and pieces.

Garage sales, and car boot sales often offer rich pickings, such as plastic toys to cannibalize. Construction sets, such as the vintage Meccano used for the south pointing chariot illustrated in the introduction, are always useful, especially as a source of pulleys, gears, and small clockwork or electric motors.

Small scale automata do not need very large bits of anything, and so, for example, offcuts of wood from more practical projects can be useful. It is worth contacting woodworkers, and kitchen installers who may be happy to pass on any offcuts.

If your ambitions are more high tech, joining a hackspace may be the way to go. The plastic walking model, illustrated in the introduction, was printed at Reading Hackspace, a peer-run community workshop. Hackspaces may offer access to 3D printing, laser cutting, CnC milling, lathes and usually many more tools, but more importantly the help and support that makers need. Details for the UK at http://hackspace.org.uk/, or go to www.hackerspaces.org for an international directory.

OPPOSITE PAGE: *The Three Graces* (detail).

BOOKS

A short annotated list of some interesting books on automata and moving toys. Some of these are out of print, but you may still find copies in libraries or on the internet.

For the history of automata a good place to start is:
Automata & Mechanical Toys Mary Hillier (Bloomsbury Books 1988).
It covers the history of automata and mechanical toys from earliest times up to toys of the 1950s and 1960s.

There are a number of interesting discussions of the cultural history of automata, mechanical men and women, and artificial intelligence, for instance:
Living Dolls: A Magical History of the Quest for Mechanical Life
    Gaby Wood (Faber and Faber 2002).
Sublime Dreams of Living Machines:
    The Automaton in the European Imagination
    Minsoo Kang (Harvard University Press, 2011).
For mechanical toys I suggest:
Mechanical Toys: How Old Toys Work
    Athelstan and Kathleen Spilhaus
    (Robert Hale 1989).
Many books aimed at collectors of toys are very frustrating for anyone interested in how they work. This one is different, with its subtitle: How Old Toys Work. Athelstan Spilhaus was a scientist, as well as amassing an extraordinarily broad ranging collection of mechanical toys.

There are relatively few general books on the traditional folk toys that I find so inspiring. A good one is:
The World of Toys Josef Kandert (Hamlyn 1992).

For wooden toys in particular a delightful book is:
Traditional Wooden Toys: Their History and How to Make Them
    Cyril Hobbins (Linden Pub. Co. 2007).

For making automata see:
Automata and Mechanical Toys Rodney Peppé (The Crowood Press 2002).

This is partly a 'how-to' book, but it also features the work of 21 contemporary British makers. There is a chapter on the origins of contemporary automata, looking at the influence of Jean Tinguely, Sam Smith and Alexander Calder.
Cabaret Mechanical Movement: Understanding Movement and Making Automata
    Aidan Lawrence Onn & Gary Alexander,
    (Cabaret Mechanical Theatre 1998).

How to Design and Make Simple Automata
    Robert Addams (Craft Education 2003).

For another approach, by a Japanese maker, look at:
Automata – Moveable Illustration
    Aquio Nishida (Fujin Seikatsusha 2002).

For paper and card cut-outs there is the wonderful:
Spooner's Moving Animals or the Zoo of Tranquility
    Paul Spooner (Virgin Books 1986).

As well as seven cut-out automata, this has a brilliant introduction making a distinction between the smooth automata of the eighteenth and nineteenth century makers, and Paul's own rough automata.

And to mention a couple more of the many available books with paper cut-out designs:
Paper Models That Move: 14 Ingenious Automata and More
    Walter Ruffler (Dover Publications 2011).
Paper Automata: Four Working Models to Cut Out and Glue Together
    Rob Ives (Tarquin Publications 1998).

WEBSITES

There is now a huge amount about automata online, with material being added constantly. Here are a few sites to set you on your way:

www.cabaret.co.uk
Cabaret Mechanical Theatre's excellent website is the place to start. Many, many people fondly remember Cabaret Mechanical Theatre at Covent Garden. Lots of good stuff and links to all over the place. Check for current permanent and temporary exhibitions.

blog.dugnorth.com
Automata maker Dug North's long-running blog is essential viewing.

www.kugelbahn.ch
This is a site devoted to rolling ball sculptures and kinetic art. It has excellent links for traditional and modern automata and for paper automata.

**For paper automata you might start with:
robives.com**
A blog, a newsletter, lots of Rob Ives designs to download.

www.cool4cats.biz
Tim Bullock's amazing cut-out designs include *Feeding Time at the Zoo*, illustrated in Chapter 2.

www.walterruffler.de
The site of the German designer and enthusiast.

Man and Marble.

INDEX